I0486273

A Production of TimothyTuohy.com

Healthcare Information Technology Integrated Project Delivery

Timothy Tuohy

Copyright

Copyright 2017

© As "Healthcare Information Technology Integrated Project Delivery"

All rights reserved by Timothy Tuohy.

No part of this book may be reproduced or retransmitted in any form or by any means, graphic, electronic, or mechanical, including photocopies, scans, recording tape, or by any data storage and retrieval system, without the express written permission of Timothy Tuohy.

First Edition

For additional information:

Go to: http://www.timtuohy.com

ISBN 978-1-387-39184-4

Cover art courtesy Jackson Health System

Dedication

This book is dedicated to my grandson. He was diagnosed with leukemia at the age of ten months, just as I began to write this book. It was this seminal event that caused me to fully understand the importance of the work we are doing in Healthcare and in Healthcare Information Technology.

The orange bar on the cover of the book and the orange color of the dialog boxes herein are all as the result of my desire to call attention to Childhood Cancer and specifically to leukemia in children.

Healthcare Information Technology Integrated Project Delivery

Table of Contents

Healthcare Information Technology In Motion

Healthcare Information Technology
Integrated Project Delivery

Preface

The hardest single thing about writing a book is how to start it. Does one start with a witty quip about the subject in an attempt to make one's self seem smart or a master of the topic. Does that make one seem smart? Or does it merely make one appear to think themselves smarter than their peers?

This book is written in haste for I fear I will not be able to get it out and therefore you will not be able to benefit from it before the process overtakes us all and changes yet again. The thing is, I have been working the Project Management part of Information Technology for fifteen years. I have studied the methodologies of the PMI, and Agile. Silly as it may seem I am one of those people who buys audio books on the subject and listens to them while I am driving. It's even worse than that! I buy the audio book and the hardcopy book and read the book while I listen to it underlining the parts I want to remember!

It is not because I think myself some kind of Project Management genius that I write this book, I find myself somewhat lacking actually. I found tools that work and want to share this with everyone who needs tools that work. I figured as I studied the topic and produced papers on the subject matter I should go ahead and encapsulate it in a book format. Maybe what I am learning can be of use to you.

If it is, maybe you will share this with your peers. As a team, maybe we can perfect this method in Healthcare Information Technology and all of us can benefit.

I work in a County Healthcare environment in Miami. Our system is large, and our primary hospital is the largest County Hospital in the country. Similar to everyone I have talked to

Healthcare Information Technology In Motion

in every other major medical system, Information Technology in our system is always under seemingly impractical time and resource constraints. I work diligently in this book to exclude people from the term "resources". The Information Technology staff as whole is overallocated and demands are driven from idealistic expectations outside the department. I believe this is due to Information Technology not being the highest technology in Healthcare, medicine is.

Planning projects in the conventional methodologies where one attempts to address the process prior to the process beginning is inherently flawed because humans are notoriously incapable of seeing the future clearly. Further, Agile methods don't apply themselves well because we are not producing original coding and have no control over the segments or sprints.

In addition to Lean Integrated Project Delivery, this book introduces the Lean Construction Institute's methodology referred to as the Last Planner System®. We have been integrating and adapting this methodology in to the Information Technology Division's projects. We too are learning this as we apply it.

Here's to continuing Healthcare Technology In Motion! No longer Project Management – let's begin Integrated Project Delivery.

About This Book

I also express my gratitude to William R. (Bill) Seed for exposing me to this Integrated Project Delivery and Target Value Delivery. The book "Transforming Design and Construction – A Framework for Change" for which he served as Executive Editor, has been the major inspiration for writing this book. This book more than any other on the reference list has been the inspiration for most of the content herein.

Healthcare Information Technology Integrated Project Delivery is not a complete treatise on the subject matter and is not intended to be. There is a list of books and authors that have formed the reference material for it's content in the References at the end of the book.

David Rockowitz, Director, IT Project Management Office, Jackson Health System and I have been working together with the team of IT project managers in our division. I started the process of writing this book in the format of a Power Point training slides so we could all "get on the same page", that Power Point Set is included in the Appendices. I will use the APA model for reference to the books, articles, web sites, and journals referenced herein. I recommend you follow-through by "going deeper" and expanding your knowledge as you need it and as we develop our skills. This book should give you the basics though, so you can begin on your own course of action.

The book will be organized similarly to Bill Seed's book as far as the arrangement of the content. Bill's book is a very good reference to own as you move forward, and I strongly recommend you find and purchase your own copy.

Like the Lean Construction Institute book for which Bill served as Executive editor, I will organize the chapters to answer specific questions:

- What?
- Why?
- How?
- When?

Healthcare Information Technology In Motion

Healthcare Information Technology
Integrated Project Delivery

The objective of this book is to provide the information and methodology we will need to implement this system, so we can stop managing projects and start delivering projects in a Lean integrated methodology.

Thanks for reading and welcome to this adaptation and application of Lean Integrated Project Delivery –

Tim

I think the single most important advancement currently occurring in the Healthcare Industry is that advancement that is occurring in Healthcare Information Technology. Until recently Healthcare IT has been focused on accuracy in billing. This is a very important part of the Healthcare Industry and I do not discount that. However, recently we are also gathering and cataloging diagnosis, treatment, procedure, and history data that will grow into the information and knowledge that will allow us (humanity) to find the root causes for such diseases as Leukemia and other cancers as well as Parkinson's and Alzheimer's and other neurological diseases. As we gather information that medical professionals and data scientists can use to find the initial indicators exponential advancements in disease eradication are possible. These diseases will be prevented and eliminated; Healthcare Information Technology is how we are going to do this. Healthcare Information Technology is contributing to the effort every day. Healthcare Information Technology Integrated Project Delivery Professionals help bring the tools into the Health Systems sooner and more effectively.

Checkout this link www.youtube.com/watch?v=ZPXCF5e1_HI

Healthcare Information Technology In Motion

1 - Healthcare Information Technology In Motion

What?

Nearly every Project Management book I've ever read begins with a dissertation of projects starting back in Egypt when they were building pyramids. This book assumes you already know that. It also makes two additional assumptions. First that you are already managing projects and know what you are doing and second that you are in Information

> *The genesis of Lean methodology stems from Toyota Production which was inspired by W. Edwards Deming and others. Integrated Project Delivery is a construction contracting methodology to align goals etcetera and the construction industry is in fact a late adopter to this concept.*
>
> Bill Seed

Technology, are customers of the Information Technology Department, are affected by Information Technology, and are dissatisfied with the approaches to Project Management you have been using.

Hopefully you can appreciate the irony of this; Information Technology Project Managers generally couldn't care less about construction. Telling Information Technology Project Managers our profession started with the construction of pyramids is ineffectual, but here I am writing a book about a methodology I was introduced to through construction.

When first exposed to this methodology my attitude was simply disgust. Here was yet another restart of projects that were already far behind schedule, some of which had not even started yet. Being the Information Technology Project Manager assigned to the major construction efforts in our Health System I was playing an observer role more than a participant role. I didn't see my role as being part of the construction team, I saw my role as defender of the allocated Information Technology budgets. Budgets that had been estimated by Information Technology executives that I would have little or no power to change, let alone increase if needed. I was being placed in yet another unwinnable scenario by people whose entire understanding of Information Technology was summed up by the words, "I.T. costs too much."

The citizens of our county voted a bond issue to contribute $830 million to our updates, renovations, and construction projects and there was a need for those citizens to see some progress. As a result, I found myself transitioning from the world of managing Information Technology infrastructure projects into the world of preconstruction. I was happy for it too. Too long the requirements of Information Technology were a total afterthought or at least relegated to a group of professionals who were not familiar with the modern scope of our profession.

Healthcare Information Technology, unlike military, education, and transportation applications of Information Technology is not typically elegant. (I refer to military, education, and transportation because these are the other experiences I possess, along with a short stint in nuclear power.) In forty years of Information Technology operational experience I can tell you the on-demand pace of Information Technology in Healthcare is unique. Unlike the centralized command and control that the other industries I have experienced possess, Healthcare is controlled by an intricate weave of physicians, nurses, and technicians. The drivers are as diverse as the diseases they fight and the needs spring up just as quickly. Doctors who make new discoveries of breakthrough healing methods can request (demand) new implementations from one day to the next. The implementation of the Affordable Care Act and the fact that there is not a "single" Information Technology platform or standard in the United States and you begin to understand the intricate weave as well as the reasons for some of the disconnect. Understanding it does not relieve the stress.

Information Technology in most organizations is the top of the technical food chain, in Healthcare the top of the technical food chain is medicine and Information Technology is just another support mechanism whose historical relevance is billing, and even that has been subjective at best. (Trotter, 2013) Healthcare Information Technology is typically understaffed and over allocated, Chief Information Officers often report to the Chief Financial Officers and many of them are treated as if they are simply high-level technicians as opposed to Information Technology leadership professionals. This is not an indictment of the political structure of Healthcare in the United States, this is an observation and I include it specifically, so the reader will understand that the writer lives in the frustrating and intensely stressful world that they live in. I assume you work in Healthcare or you would not be wasting your time with this book.

Why is it important to understand this? The world of the Healthcare Information Technology Project Manager is more on-demand than any other I am aware of. Most projects are presented as "just get it done" affairs and they are given with deadlines that are ripped from the air as if they can be completed by taking an oral medication. The level of planning that is typically allowed for projects in other environments is simply not offered in Healthcare Information Technology.

Electronic Medical Records and the Affordable Care Act along with HIPAA and the HITECH Act and things like Meaningful Use are driving Healthcare into the twenty-first century and establishing Information Technology as a center piece of Healthcare. Physicians and nurses now must use Information Technology as much as they use a stethoscope, it is no longer an option to leave documentation for later or to enter free style notes in a paper medical record or chart. Heath Care Information Technology Project Managers spend most of their time working to implement these Electronic Medical Record software components and modules as rapidly as possible and the elements thereof are medically technical in the extreme. They find themselves working with Information Technology employed Registered Nurses and Physicians in an environment that would be alien to Information Technology in any other industry.

Projects managed in the conventional manner of working to produce a plan for something to occur in the future through the four steps; Planning, Build-up, Implementation, and Close Out (Harvard Business Review, 2012) simply don't work. We apply these methods because we need tools and there are management demands for status and progress reports, but the project manager is relegated to being a 'box checker' at best and a task master, similar to what I imagine were employed in the construction of the pyramids, at worst.

Healthcare Information Technology In Motion

Healthcare Information Technology
Integrated Project Delivery

Projects managed in the agile methodology generally are not even considered. Even though the rapid, iterative application of installing and integrating components and modules of the Electronic Medical Record may make one 'feel' like an agile methodology might work. The Agile Manifesto (Agile Alliance, 2017) and Twelve Principles are organized specifically for developing and delivering software. Most Healthcare Information Technology organizations are not developing software but are implementing someone else's software and either customizing it to their organization or forcing their organization to comply with the original structure. In the case of the Health System for which I work that Electronic Medical Record and the associated systems is provided by Cerner. (Cerner, 2017) Because of this, agile methodology is not applicable in a meaningful way in the Healthcare Information Technology environment, at least, not in ours.

Projects managed in construction apparently experienced similar problems. Like Information Technology, the construction industry was seeing projects exceed the planned time and cost, as well as creeping scopes and customer dissatisfaction. Their solution was to look to the methodologies that are defined by the Lean Principles that have grown up through the years from 1913 when Henry Ford implemented the first Lean model to his entire production line. No one called it Lean back then but that is what it was. More recently Kiichiro Toyoda and Taiichi Ohno shifted the focus at Toyota after World War II to the model widely known today as the Toyota Production System (Liker, 2012). Lean techniques were already in practice by Eiji Toyoda and Taiichi Ohno when the end of World War II thrust upon Toyota postwar Japan's fertile environment for new thinking and W. Edwards Deming's quality enhancing

Healthcare Information Technology In Motion

11

ideas were adopted. (Womack J. P., 2007)

Projects managed in the Lean Construction Institute's Last Planner System® of Integrated Project Delivery introduced the construction industry to a new version of project management. Target Value Design is the methodology used to set project targets and steer design and construction toward those targets. This process was adapted from the target costing product manufacturer's methods. Through the years from 1996 until 2007 there has been a series of progressive iterations and steps that have lead up to the codification of Target Value Design (Ballard I. D., 2016) by the Lean Construction Institute.

It is this methodology that I am adapting and introducing to you as Healthcare Information Technology Integrated Project Delivery. There is no more chaotic environment in Information Technology today than the one that has been thrust upon it in Healthcare. Admittedly, not as chaotic as postwar Japan.

Why?

The reasons can be delineated all day, but if we apply Occam's Razor, all things being equal, the simplest explanation is probably the truth. The simplest explanation for why is, because the current methods aren't working. We are working harder than we have ever worked before and falling further and further behind. Working attitudes continue to slump despite the best efforts of the leadership, not because the leaders are doing something incorrectly, but rather because they have no control over the tectonic changes that are being thrust upon the world of Healthcare. Complicating the tsunami of regulatory requirements flooding over the business of Healthcare - technology is changing at an astounding rate as well. We tend to think of these tectonic changes as individual events, but they are not! The regulations seek to force Healthcare into the twenty first century by compelling compliance with technological record keeping but the technology is progressing as well - making the targets increasingly difficult to hit.

Since their inception in the 1980's by Motorola, cell phones have done more to change our society that almost any other technology. You may not remember "Simon" - introduced by IBM and BellSouth in 1993 but I bet you remember the PDA – the Palm Pilot was a hand-held computer. I wrote an entire book on Dell's version which had a Windows operating system and would allow easy transfer from mobile Word application to my desktop computer's Word application. The true tectonic shift though began when Apple introduced the iPhone in 2007 combining the phone with the PDA. (Pothitos, 2016) Now, there is no escaping these devices.

Let's overlap the laws and you'll see why I am writing that we are caught in a perfect storm of tectonic changes affecting Healthcare Information Technologies. HIPAA was signed into law in 1996, there was no interchange of data between Healthcare practitioners at all at the time. Purists will argue medical records were for the most part paper, but this book is not designed to give the deep history of either the technology or the law. The HITECH Act was signed in 2007, followed by the Affordable Care Act in 2010.

Cerner was founded in 1979, in 1994 they had only about 30 clients. In 1997, they introduced their flagship Electronic

Healthcare Information Technology In Motion

Medical Record, Millennium. By 2005 Cerner had grown to more than $1 billion dollars.

Like Toyota in postwar Japan Information Technology in the Healthcare environment is embroiled in what could be called chaos or could be viewed as a fertile environment for new thinking.

How?

This book is designed to walk you through the steps quickly and easily; a sort of **"Vade Mecum"**. Vade mecum is Latin for "go with me." In English it has been used since the 1600's as a term for manuals or guidebooks sufficiently compact to be carried in a deep pocket. But from the beginning, it has also been used for such constant companions as gold, medications, and memorized gems of wisdom. (Webster Dictionary, 2017) "Go with me" is very much a methodology of the Toyota Way and therefore fits the spirit of this methodology, it is as valuable as gold, and as important as your medications.

Figure 1 – An early effort to apply Last Planner System

Healthcare Information Technology
Integrated Project Delivery

When?

When is – "Now is a good time?" – Anthony Robbins.

It is not too early to start learning the system. Don't worry about doing it correctly right from the start. Lean Learning is about continuous improvement!

2 - Engage Deeply

What?

Omnino Collaborare. Labor together. Latin from *com* +
laborare.

Collaborate – REALLY COLLABORATE with the objective
of truly, constantly, and incrementally improving the process
and delivery of projects.

Most Lean process promoting writers tend toward the
Japanese words because so much is based on the work done
by Eiji Toyoda through the years. An American self-help
coach, Anthony Robbins, coined a word "CANI!" – really an
acronym for
"Constant And
Never-Ending
Improvement!" so
he wouldn't have to
use the Japanese
words "Kai Zen"

> Semper Melius or Kaizen …
> The Japanese word is widely
> used and may be easier to
> remember – but I like the Latin

which mean change and good. Together "kaizen" is the term
used frequently to refer to the process of finding and making
small incremental improvements in pursuit of perfection.

I like to use Latin because we use it so much in the medical
world, in the case of continuous improvement, Latin tends
toward "semper melius." I like it better, but most Lean
thinkers prefer to honor the Japanese words.

What is it about "semper melius" that applies itself to this
topic? Collaboration, should include all the stakeholders of
the project, is a systematic study in pursuit of the perfect
delivery of the project's product or outcome.

Healthcare Information Technology Integrated Project Delivery

What does this mean? Collaborate – really collaborate.

This means as Healthcare Information Technology Project Managers we will need to broaden our horizons. We need to include not only the Information Technology Department nurses, physicians, programmers, technical support, and infrastructure staff but also the end users. Our goal is to bring *all* the parties together in collaboration from the beginning to establish true working understanding of the needs of the Health System that we as Project Managers are trying to fulfill.

I know, right? This is going to be fun!

Why?

Managing Projects in Healthcare Information Technology has not been considered as a process that can add value to the total beyond the thought "the sooner the software or product is in service the better we can perform patient care." This is because the sooner we can get the electronic medical record to be the single source of truth for the patient the more likely we are to provide better informed care. There is however, a Healthcare Information Technology aspect that has a much more profound legal and financial implication. One of the primary Stage 1 Meaningful Use criteria is consistent with Medicare and Medicaid law focusing on capturing health information in a structured format. (Trotter, 2013) Most of what we do as Information Technology professionals in Healthcare is work toward a "true north" that there is value in "normalized" or structured data. Ultimately, the 'big drive' is to stabilize the methodology for billing, so that payers can have some level of confidence that they are not paying for a plethora of unneeded activities that pump millions of dollars to people who are performing unnecessary procedures to line their own pockets. (Uhlman, 2012)

There is an important financial and legal reason why Information Technology Project Managers need to find a more effective, efficient, and expedient way to successfully deliver projects. The success of our entire health system depends on it. The primary purpose for collaboration across all the parties both providing and affected by the project, is to establish the definition of "true north" as it applies to the project being delivered. (Lean Enterprise Institute, 2017) Having a "true north" allows us all – as a team – to come back to the right direction to accomplish our goal. If the path we are taking is not true north, then we can quickly and easily adjust to the course that will take us in the direction of true north.

The biggest reason why is this: Without deep collaboration with all the associated people related to the project, there are hurricane category winds that will blow us off course seemingly on a whim. By engaging deeply with all those associated with the delivery of the project, and allowing them to own it, we as the Integrated Project Delivery Professionals are no longer the 'task master'. We have transitioned to Project Facilitator. It is a very agile concept for a scrum master to be the one who knocks down obstacles for those people who are occupied producing working software; in Integrated Project Delivery methodology a similar transition occurs. It is at once liberating and fulfilling for the Project Manager.

How?

Use a Big Room Approach. For this chapter, we want to focus on some basics to successfully building collaboration teams. We have been working with this methodology for only a short time, and I can tell you honestly, I am still a student. You might have picked up on that by the construction of the 'chapters' of this book. You'll note that each chapter is written much like a school paper, broken down into sections and providing references to the material. From this you can derive that I am studying the material as I write and practicing it in the lab of real-life applications. Integrated Project Delivery is a transformational change.

This is the first part of how. We're Project Managers, we may even be Information Technology professionals. Some of us may have Healthcare Administration degrees, but we are not surgeons, doctors, or clinicians. I have worked with one Healthcare Information Technology Project Manager that had a Nursing Degree, but she moved on to "better things". The first thing we must do is be humble. We must free ourselves from the notion that Information Technology is the solution and we are the providers of such. (Uhlman, 2012)

We will approach the team with the attitude that we are working to learn what is needed from each other. Respect for the people we are working with is the corner stone of success. Focus on their definitions of value as it applies to the project delivery. What is important to them? Remember, being both the Project Manager and a technical expert or Subject Matter Expert can work against you because your team may 'let' you handle the stuff you appear to be a master of. However, we should also become increasingly proficient in the tasks we have to perform, people listen to us more seriously if we are the experts. (Portny, 2013)

Next, focus on driving out waste. We are still managing a project, we are not there for a group complaint session, we are there to collaborate. We are collaborating to deliver the project on time, under budget, with the resources provided. Collaborating is the single best way to drive out waste if everyone on the team is striving to do so. Focus on eliminating steps that are duplicated or are unnecessary. Emphasize eliminating wait times and on working parallel efforts. Place attention on work happening in the team member's respective specialty that can be leveraged to get this project delivered. (Glenn Ballard, 2016)

Continuous improvement of processes is a key element of this adaptation of Lean Integrated Project Delivery. Continually ask, "How can we do this better, faster, more cost effectively, and with less work effort?" Semper Melius.

The group, the team, grows always more faithful and trustworthy to accomplish the goals of the project because they are working together as a team each knowing all the challenges each of the members of the team faces. We as Integrated Project Delivery Professionals now enable the team members to succeed.

> Semper Fidelis. Always faithful.
> Semper ad Meliora. Always toward better things.

When?

At the beginning. As soon as we are assigned the project we should begin asking with whom do we collaborate. There is a problem-solving methodology known as "the five whys" – practice this methodology when selecting the "who" to collaborate with. Who asked for the project? Why? Was there someone else asking them? Who? Why? Who will be affected? What technologies will be employed? (Ohno, 1988)

In short, when we begin a project we **begin** with collaboration because we are not the ones who are going to use the outcome, product, or service that results from our project. We are not the subject matter experts, we are not the medical professionals, and we are not doing the actual work to deliver the project. We are Integrated Project Delivery Professionals.

3 - Collaboratively Plan

What?

E Pluribus Unum – from many one.

Respect for people. Its basic really, and it is the basic building block for this methodology. Consistent with both Agile and Lean principles our goal is to develop a culture of respect and continuous improvement. (LCI, 2017) (Agile Alliance, 2017) As Integrated Project Delivery Professionals

we strive to build respect for people based on team building and commitment stability. We can depend on each other to make and keep commitments because we are making these commitments to each other and continuously working to improve our ability to deliver on the commitments we make. It is important to us to learn as we grow. Initially our commitments are less accurate than they are after a cycle of iterative processes of planning and committing. The more iterations we cycle through the more accurate our plans become. As we develop our skills for planning, committing, and delivering we become more confident, not only with ourselves, but also with the other members of our team.

This is how we develop trust. Gradus per gradus – step by step. As each of us commits to continuously improve our delivery commitments, and accept the delivery of those commitments from the team members we have received commitments from – and the consistency of the deliveries increases – trust grows within the team. Through this process of constant improvement (semper melius or kaizen) team members begin to feel increased pride in their own work positions as well as finding satisfaction with the team. We desire that every member of the team feels they are as important as every other member and that the team is the most important thing in the project process second only to the delivery of the project which is the purpose of the team. We are consciously working to reach the best possible resolution of the top three levels of Maslow's Hierarchy of Needs. (Fowler, 2014)

This is aimed at creating more value for our customers and stakeholders through Lean Integrated Project Delivery. The Lean Integrated Project Delivery System is an organized implementation of Lean Principles and Tools. Together the team can employ these to function in unison to deliver the project. (NASA, 2017)

Why?
Ad Perfectum – toward perfection. Humans are notoriously poor planners. Our forward vision is great as it applies to a goal or ultimate objective, such as built buildings or climbed mountains, but the process of planning all the individual steps necessary to actually build the building or climb the mountain

are seldom accurate. No matter how many networks we have built, buildings we have built, or mountains we have climbed, each network, building, and mountain is different. In Healthcare Information Technology every project is somewhat different from every other one because the stakeholders are different, the desired outcomes are different and without fail the function we as project managers are supporting is literally life and death. The outcomes we desire in Information Technology are accurate, usable, data that can inform knowledgeable decisions by clinicians and researchers to heal and prevent disease, wounds, and damage. In our case as Healthcare Information Technology Integrated Project Delivery Professionals this includes a complex weave of infrastructure, software, hardware, interfaces, and methodologies. It is impossible for a single person to understand all the various aspects and complexities of the system. To become more nearly perfect with each iteration requires that we collaboratively plan. To do this we include all team members to the level of the last planner. The Last Planner® can be the last person in the chain of command that has the responsibility of planning the implementation of the product of our project. This can be the Team Lead in the Network Infrastructure group or the Nurse Manager. (Glenn Ballard, 2016) (Ortiz, 2017)

How?

Trahere concilio – Pull Planning. Pull planning has two primary origins. The most commonly understood seems to be the idea of a 'virtual project retrospective'. The term retrospective is often used in a post-mortem thought process. In this methodology, after the project is complete, the team gets together and reviews all the things that happened both good and bad to determine what could have been improved upon, what went well, and what failed and must be reinvented before the team attempts another project. A virtual project retrospective could be referred to as a 'pre-mortem'. Imagining the project is complete or the milestone is reached and then imagining all those things as if they had already happened.

The other, and the one I like better, is the concept of pulling a part from the shelf in an auto parts facility in a car dealership. The part can be anything really, from a can of paint in a warehouse home improvement store to a pack of rice in a grocery store. The auto part is best in my mind as an example. Let's examine a simple part and pull-plan backwards. (Womack J. , 2003)

Let's use a headlight, simple enough, right? So, imagine you are standing in front of the shelf. The headlight is exactly where it should be, in the packaging it should be in, properly labeled, priced, and available. This is the state the product needs to be in to be usable by you as you fill customer orders. The milestone is 'parts in stock for sale'. It is the last step of the process that allows you to open your parts store to the customer.

What had to occur for the packaging to be correctly printed, shaped, cut and formed? Thinking backwards from the finished product, what was the last function to occur? You are correct if you suggest that was placing the light in it. Prior to the light placement, the packaging was folded into a box. Who or what folded the box? Who or what designed the cut of the box? Who designed the print format of the box, wordsmithed the verbiage on the box, and added the logos? Was the box printed in a flat format and cut? Remember you're thinking backwards from the finished product.

Healthcare Information Technology In Motion

In either case, this process is framed by the questions: "What was the last thing done to make this milestone complete? Who did this? How long did it take?"

We're going to spend the next few chapters developing this methodology, but it is important now for us to start thinking who do we need to collaboratively plan with in order to reach the milestone of 'usable product', which in our case is information.

> *"The first step is transformation of the individual. This transformation is discontinuous. It comes from understanding of the system of profound knowledge. The individual, transformed, will perceive new meaning to his life, to events, to numbers, to interactions between people."*
> W. Edwards Deming

When?

Nunc est tempus – Now is the time. Here we are reading this thinking when do we get this collaborative team together? When is our project supposed to start? When is it supposed to be complete? In the Healthcare Information Technology environment we work in, the best answer is something like "yesterday"!

As frustrating as that is, it is important that every day that passes without Information Technology fully supporting the medical services is another day when doctors and nurses are making decisions based on incomplete data and insufficient knowledge. This is not saying the doctors and nurses are not knowledgeable, it is saying we are all working daily to get better, to learn more, to be more accurate in our assessments and diagnoses. Data gathered is the seed that germinates into information as it is stored properly in databases that are usable, but the fruit of that information technology is knowledge.

Now is the time.

4 - Find True North

What?

Imagine you are flying an aircraft, even if all you are doing is practicing or site seeing, you must by necessity know where you intend to land. For the purposes of this book imagine that you are flying from place "A" northward to place "B" in a straight line. As you become airborne you remember for the entire trip you will have a crosswind from one side. You are also aware the crosswind is gusty rather than steady, so you will be unable to maintain a simple adjustment but will need to adjust frequently to varying amounts or you will be blown off course. If you were only traveling a few hundred feet this would be no problem, you would be able to see your target runway and adjust to it rapidly. However, for the purposes of this example you are traveling for hundreds of miles and the wind intensity is sometimes severe.

If you do not keep track of your position relative to your planned course and adjust back to it, you will soon be lost and have no idea where your intended landing zone is.

Healthcare Information Technology Projects are often similar to this example. We start off with an understanding of what we are managing to but along the way there are winds of change blowing us about and in many cases, we have no way of pushing the project back on course. Instead, we spend hours and hours attempting to salvage the intended goal and just somehow get the project online or live and get out of it with our sanity.

The concept of "true north" (Ammer, 2003) is an American idiom that originated in the late 1900's that refers to the concept that we must have an absolute purpose and focus

Healthcare Information Technology
Integrated Project Delivery

toward which we are steering. Without this, we are subject to meandering off the path and becoming lost in the "Muda" (waste) of lost time and effort. (Lean Enterprise Institute, 2017) Worse yet would be to become so off course that we are lost in the waste land – vastum terram.

The number one reason for project failure in Healthcare Information Technology is scope creep and uncontrolled change. It is notably one of the top three reasons all projects fail. (Project Management Institute, 2013)

Why?

If you don't know where you want to go, then it doesn't matter which path you take. So said the Cheshire Cat in Lewis Carroll's classic book. (Dodson, 1865) This is an absolute in the world of Healthcare Information Technology Integrated Project Delivery. Defining **true north** at the outset of the project gives us a much better chance of delivering the project on time. Otherwise, every time we come to a fork in the road we are taking a chance our choice may lead to a failure to deliver the project on budget or on time or at all.

This is not a new concept, it has been taught since earliest Greek Olympics, when the concept of laying off all the things that weighed a runner down led to their competing naked. Similarly, sculptors cut away anything that doesn't look like their subject.

In more recent times, Toyota implemented the idea that all waste in the production system should be cut away. The basis being the least amount of effort needed to produce a quality vehicle was the best method of manufacturing. (Ohno, 1988)

Healthcare Information Technology In Motion

29

Healthcare Information Technology Integrated Project Delivery

How?

Up until now all the chapters in this book have been leading us to this point. This particular "how" has been founded by all the things written in this book so far. Now we are challenged as leaders and coaches to bring the integration of teams into alignment helping to coach and guide them to a lean project delivery that includes their various views and desires. By following The Last Planner System® methodologies that we will learn together as we progress in our profession.

Use the Big Room concept and apply the principles of this system to develop a **Conditions of Satisfaction** document. This document is not a scope document as you might have tried to develop before but is more like a list of conditions that are expected by the end users developed collaboratively.

"This is an explicate description formed by the end users of actual requirements that must be satisfied in order that they may feel that they have received exactly what they wanted." (Kristen Hill, 2016)

Collaborate deeply with all the parties. How often have we started a project thinking we understand what the end user or chief stakeholder wanted but once we got involved we learned there was some major element missing for the implementation that was going to require a lot of time and a lot of effort to resolve?

Let's use the following example, an existing busy clinic is chosen to open an Urgent Care Center. It is an executive decision and makes perfect sense. However, when the workers begin to make the modifications they discover the facility is in considerably worse condition than expected and the work will cost much more and take much longer to complete. At the outset the parties were not collaboratively engaged. The existing managers knew the condition of their clinic, they could have helped inform the construction people and architects, who in turn might have been able to better inform the management as to the scope of work and cost. By keeping all the parties deeply engaged the project manager could have saved the effort of having to "walk it back" several times as the onion being peeled continued to find more and more difficult circumstances.

The parties were not engaged so the project took much longer and cost much more than was originally planned for. Using the methodology we are learning in this book we might

> The ultimate True North is to collaborate, to build individual commitments, between the individuals who are working to deliver the project. These individuals own the project collaboratively and deliver the project as a team. Coaching to this outcome is the purpose of Healthcare Information Technology Integrated Project Delivery Professionals.

have discovered some of these issues in advance and built in contingencies and escape plans. Further, the Information Technology processes were not addressed above the network cabling, and security was completely overlooked.

In this example the Information Technology Project Manager did everything correctly as it applies to the common methods of project management. These are examples of what we seek to learn from and improve upon. We know these types of efforts will continue because we are in healthcare and this is

Healthcare Information Technology In Motion

how it happens in healthcare, but we can improve and learn by applying the methodologies laid out in this book.

When?

It is not too late to start looking at the compass and working out what is true north in our projects. As a team we will have to continue to work toward the Integrated Project Delivery model. This means that starting now, we, as Healthcare Information Technology Integrated Project Delivery Professionals, need to start working together to see where our assigned projects interconnect with the overall target value of the organization.

Does the Mission Statement of our organization give us some level of understanding of our health system's true north? How can we integrate all the Information Technology projects into this methodology?

Now is the time to find and define the Conditions of Satisfaction. Now is the time to find out what is really wanted and the best way to do that is to engage everyone from the end users, doctors, and nurses, to the infrastructure support team, to the coders, to the installers. Who knows if the Conditions of Satisfaction would have changed in the example above if all the parties had discussed everything that was known as well as everything that would have to be done to reach those Conditions of Satisfaction? Maybe the Conditions of Satisfaction would have been negotiated into a different set of Conditions or maybe the requirements of Satisfaction may have changed.

Now is the time for us to think about all the various individuals who are working sometimes unseen by the work flow but whose efforts and the things they build, or support

Healthcare Information Technology
Integrated Project Delivery

are directly influential to the implementation of or outcome of the project. Let's include them now.

Who are the "last planners" in our project?

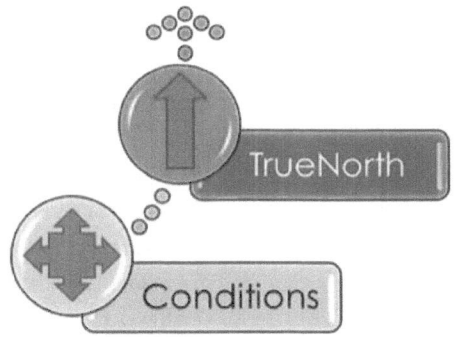

5 - Last Planner System®

What?

The term Last Planner is a registered trade mark of the Lean Construction Institute (LCI). In this volume we do not adhere strictly to the methodology of the LCI system but are heavily influenced by it. (LCI, 2017) The goal of the system is to bring stability to the project by giving attention to flow while reducing waste created by the transition of work from one specialty to the next thereby continuously improving the project workflow situation. (Seed, 2017) The Last Planner System of Production Control® was developed by Herman Glenn Ballard in his thesis submitted to the Faculty of Engineering of the University of Birmingham for the degree of Doctor of Philosophy, May 2000. (Ballard H. G., 2000)

In his thesis Doctor Ballard states, "This thesis extends (The Last Planner) system application to those coordinating specialists, both in design and construction, ... one of which also explores the limits on unilateral implementation by specialists."

Why?

The Last Planner System® (LPS) produces and expands predictability of project workflows. Results are the outcome of people working together in iterations of planning discussions that produce a "network of commitments" essential to identify the work, tasks, and products needed to accomplish the project milestones. This network of commitments makes work output (product) ready, and assures a **person** has promised to complete it. All the while, we are learning from our experiences gained in the iterative process. This must be accomplished in a collaborative environment. Commitment and responsibility are two-way agreements!

This network of personal commitments, not a published schedule or project plan, is the supporting instrument or tool, for engaging the project team in completing work consistent with the deliverables promised for the project. (Greggory Howell, 2005)

The Last Planner System® can be tailored to project circumstances, and we find it very helpful in Healthcare Information Technology projects because the intention of the system and the fundamental nature of the practices involved are clear: "Produce predictable uninterrupted workflow by creating a coherent set of commitments that connects the work of the specialists to the promise of the project to the client and coordinates their actions."

Healthcare Information Technology In Motion

Healthcare Information Technology
Integrated Project Delivery

This happens in five iterative conversations designed so the team can manage the network of commitments inside each of their accountabilities and specialties.

How?

Trahere concilio – **Pull Planning.** As the Healthcare Information Technology Integrated Project Delivery Professionals, we facilitate the team of people who are ultimately responsible for delivering a project milestone. We help them plan together how that will happen. The planning conversation produces an integrated project delivery plan for what will be delivered at each hand-off. The hand-offs are defined by the team members during the Pull Planning meeting or session or conversation, and the output of that conversation is a schedule for those hand-offs. This schedule prepares the team for action together. The future is not set, and remains uncertain. The plan developed is based on a series of requests starting from the milestone and working backwards – **the pull**. Each of these requests form the basis of the individually identified and personally promised tasks or deliverables that translate into the network of commitments necessary to deliver the phase and establishes when the work should happen.

There can be two conversations if the project is large enough. One of these would be to pull the various milestones of a larger project. And then the pull planning sessions can be performed to individual milestones. This chapter only deals with pulling to a single milestone.

The Healthcare Information Technology Integrated Project Delivery Professional facilitates this conversation but the actual tasks, durations, sequence of events, and assignments are accomplished by the team members themselves. The team is making promises to each other, not to the Project Manager. This is completed in the next conversation, but often, due to the time constraints imposed on Healthcare Information Technology is actually performed in the Pull Planning conversation session.

Promptus ut congue - **Make-Ready Planning**. Each team member has graduated from interested observer to responsible individual for a group of actors (crews, teams, and individual persona) and reviews the work in the coming six weeks look-ahead period. She or he identifies whatever is needed to accomplish the work and makes the requests and receives promises (personal commitments) needed to assure the input or requirement will be available when essential. Most (if not all) of these tasks will have been identified in the pull plan, minus the durations and sequences. In this conversation personas ask for help whenever they lose confidence that the work will be ready when required. They know their part of the work but there are other workstreams that must also be working to create a Lean Workflow.

The team adjusts or reaffirms the plan in regular meetings during the lookahead period. No single individual is responsible to the whole plan, but every single individual is responsible to the team to accomplish their part of the plan. The make-ready plan establishes what can happen and serves as the basis for securing reliable commitments for the coming scheduled work. The commitments are made from the last planner in each specialty to other last planners in other specialties. This is where The Last Planner System of Production Control® derives its name.

Consilium hoc septimana - **Weekly Work Planning.** Final coordination of the work in the coming week (or two) is completed as each last planner makes promises to the other last planners and the project manager asserting what will be delivered, to whom, and by when. Capacity, resources, and finances are allocated by day in support of those promises. These conversations promising deliverables occur in a group setting. This meeting allows other last planners and team members to assess consistency with their own promising. As a result, adjustments to the weekly work plan (WWP) are made during the meeting.

Promissa dies gerant - **Daily Commitment Management**. Like its cousin the SCRUM meeting, (Layton, 2012) this is a formal (not a hall meeting) however brief "stand-up" meeting (or conference call). The appropriate members of the team are involved and are held daily to report WWP completions. During these meetings the team lead, or last planner works with the team to adjust to the circumstances, re-promises as necessary to the project team, (through the Project Manager). Personas also get help from each other and record plan variances and their reasons. This is the root work meeting and serves a self-directed controlling or steering purpose that allows those closest to the work to adjust to the always-changing conditions of the project, so they can complete all the ir promises. (Greggory Howell, 2005)

The Last Planner System® drives planning reliability to continually improving levels.

When?

As soon as we know there is a project, we should start assembling the team of last planners and stakeholders. We should include everyone who contributes to, has a deliverable in, finds value from, and is affected by our project when we are seeking with whom we should collaborate in our project.

Healthcare Information Technology
Integrated Project Delivery

When we have the definition of the final product we should begin planning retrospectively from it. That definition is our **true north**.

Right now, we should start reprogramming our minds from the conventional waterfall project planning forward methodology to that of The Last Planner System®!

6 – Milestone Planning

What?

I'm calling this milestone planning however in The Last Planner System® (LCI, 2017) it is referred to as Master Planning. Our objective in this step is to build the Conditions of Satisfaction and identify the milestone that led up to those conditions being fulfilled. To do that we need a spectrum of team members ranging from the doctors and nurses and technicians that will be end users to the technical consultants that will configure the software and the infrastructure technical managers that will determine what additional hardware, operating systems, and network devices are required. (Kristen Hill, 2016)

This meeting should be held in a "Big Room". A Big Room is a conference room capable of supporting 25 to 30 people with available large whiteboards and space for **pull planning**. In the section on "how" below, we will examine the physical processes of pull planning in a Big Room environment.

From the executive level and that level that includes the doctors, as well as from the user level that includes the nurses and technicians, we are approaching the milestone effort as if we were filling out a "dream sheet". This means, we are looking for what they think is the best possible outcome of this desired project. In their world, what is the problem to be solved, the enhancement to be installed, the workflow to be improved, or the work effort to be eliminated.

The answers to these questions constitute the preliminary Conditions of Satisfaction.

Healthcare Information Technology
Integrated Project Delivery

Next, we begin to pull backwards from this ideal set of Conditions of Satisfaction, in a virtual retrospective, walking backwards through time, along with the technical experts that will be managing the work required to accomplish the desired outcome. As we begin to perform this exercise, certain discoveries begin to be revealed. As an example, it may be revealed that this outcome will require additional workstations, these workstations may need additional computing power as compared to the existing workstations; the existing workstations may need to be upgraded; additional network jacks may need to be installed, and additional network switches may need to be ordered to support the additional ports required. Servers may need to be ordered and time will need to be spent determining what is the optimal operating system for the servers. What additional support staff will be required, both in the desktop the server and the software specialties?

The purpose of this meeting is to identify milestones not to identify costs or difficulties. What milestones need to be identified to reach the optimal Conditions of Satisfaction set forward in the dream sheet.

In another iteration of the same conversation high-level Subject Matter Expert opinions may be discussed among the chief stakeholders and end users to perform "Not to Exceed" and / or "Rough Order of Magnitude" estimates and reach a determination of what is the maximum budgetary limit the owner is willing to spend to reach the optimal Conditions of Satisfaction. In this iteration it may be determined that if the cost exceeds a certain level the Conditions of Satisfaction would be altered or diminished to provide a working model of a product that is satisfactory to the stakeholders and end-users that may not reach the highest level of their original desire.

Depending on the size of the project this meeting may be required multiple times; but it is important that this meeting occurs and that you have a plan of what the final milestone and all the milestones leading up to it are. Each of the milestones should be separated by the discipline area or specialty that will be performing the work. These separations form lanes of responsibility are referred to as "specialty swim lanes", and we will discuss that in the section labeled "how" below. (Seed, 2017)

Why?

There are advantages in this practice. One of the primary advantages of this practice is allowing the executives and end-users, those people who will own the product once the project is complete, to understand both the work effort and the cost of accomplishing their dream. An ultimate milestone described by the person that desires it without complete comprehension of the cost or work effort required to accomplish that outcome is best described as a dream. In all practical terms the dream is a good thing. There is no rule that specifically states that a dream is vaporware. Without a dream, without a vision, there is no advancement in healthcare technology. We as Healthcare Information Technology Integrated Project Delivery Professionals thrive upon producing the miracles that end up making dreams happen.

This iterative meeting or "conversation" is not designed to talk the dreamer out of their dream. Rather, it is designed so that all the participating management parties may have input into the milestones that will have to be accomplished in order for that dream to be realized. The realization of that dream may require innovation, imagination, and new design. Through the iterations of this conversation it may be determined that the dream is entirely reachable sooner, with less effort, and less expense. This conversation may also reveal that the expected work and expected cost are exactly what the dreamer expects. Of course, it could go the exact opposite way to be determined that this is much costlier, take much longer to complete, and may not even be a realistic dream. We may not have the technology to accomplish it.

This is a conversation where we are building the document that is the description of the specific expectations of the stakeholders as to what are their Conditions of Satisfaction.

How?

Let's get down to nuts and bolts. While sitting around the table in the big room we, as the Integrated Project Delivery Professionals, realize our new role. Now our role is that of facilitator. Facilitators, in this conversation ask questions. So, in the first iteration, we may begin asking questions like:

- "What will this look like?"
- "How will this be triggered?"
- "What data will it pull up?"
- "What do you need that data to look like?"
- "How do you want that data to be updated?"
- "When do want that data to be updated?"
- "Who will have access to that data?"
- "What reports will be needed?"
- "What size monitor will it be displayed on?"
- "Where will the monitor be?"

Healthcare Information Technology Integrated Project Delivery

- "Who will have access to it?"
- "Where will the data be stored?"

You get the idea. There are many more questions that we as Healthcare Information Technology Integrated Project Delivery Professionals have learned to ask through our own experiences. The difference is we DO NOT DESIRE to be the expert in this conversation. We are approaching this to help our clients answer a simple set of questions:

- What will you give?
- What do you need to be given?

Even if we know the answers we pretend to be ignorant. Because we want the individuals in the room to personally commit to each other. We are facilitating **their work** effort, not doing the work for them.

You will notice, that the question of when is not included in this list. We will come to that question as we **pull** this plan from the milestone backward in time to its inception.

We, as Healthcare Information Technology Integrated Project Delivery Professionals are looking for a complete description of what the outcome ... what that <u>final milestone</u> looks like.

In this first iteration, we are not looking for cost. We are not looking for how much technology is involved. We are simply looking for what that primary stakeholder and those managers of the end-users think they want. We are building a final milestone. That milestone is still a roughhewn stone, it does not reflect the end result that will describe the final milestone.

Healthcare Information Technology In Motion

Healthcare Information Technology Integrated Project Delivery

Take that first description, and draw diamond on the whiteboard on the far-right side. In that diamond write something that describes that dream. It may only be the name of the project and the words "go live".

Now comes the fun part! Get everybody up out of their chairs and have them stand at the board with you. Provide each of the associated parties with a 3 x 5 for a 4 x 6 sticky note pad on which they can begin to write the milestones that add up to the completion of the final milestone. (We need to bring some felt tip markers to the meeting as well.) It is very important that we clearly explain that we're not looking for the tasks that are required to accomplish the milestone but rather the <u>milestones</u> for each of their individual disciplines and specialties that must be accomplished in order for the dream milestone to 'go live'. So, if in order for this new system to go live, the operations team must supply fifty desktop computers of a certain configuration that must be running a specific operating system and must have certain software installed upon them there are a number of tasks that you are aware of in that discipline that must happen but what is the milestone that we put on the board? This milestone may be "Installed workstations."

Additionally, this new system may need to new network switches installed in different closets because the additional workstations and the noncontiguous locations require two different switches in two different closets on two different floors. Similarly, additional servers may be needed in the data center, and that may drive a need for additional power. There may be additional need for software licenses. But the milestones associated with this effort are:

- "Installed switches"

Healthcare Information Technology In Motion

- "installed infrastructure wiring."
- "Installed servers."

In this iterative conversation we are not looking for the duration of time that is required to get the job done. That's done in the next set of conversations. The **pull plan** that pulls the tasks backwards from the milestones also adds durations to the timeline.

If your project requires a specific end date now is the time for the owner to specify when that date is. Hopefully this iterative exercise has given enough high-level decision-support material for the future owner to understand the scope of the work and therefore be willing to negotiate a more realistic desired due date.

From all these questions and all this work, we as Healthcare Information Technology Integrated Project Delivery Professionals are looking for answers to the question:

"What are the Conditions of Satisfaction?"

This tells us the answer to the underlying question that every Healthcare Information Technology Integrated Project Delivery Professional asks of themselves when they start a project, "how do I know when I'm done?"

When?

The set of meetings, even though it could be just one meeting, should be considered the **project kickoff meeting**, if you're thinking in the old manner of project management. What we as Healthcare Information Technology Integrated Project Delivery Professionals are purposing is to completely change how we have been approaching projects. As Healthcare Information Technology Integrated Project

Healthcare Information Technology
Integrated Project Delivery

Delivery Professionals, we are facilitating the teams of people, individuals, who actually do the work in accomplishing their own projects. We, as Integrated Project Delivery Professionals are helping them fully understand the work that they and their teams will need to perform.

One of the key processes in these iterative **pull planning** session meetings is to begin forming personal relationships between the individual performers. We are doing that by getting these individuals to individually commit to each step of the process that must be accomplished, not to us but to the individual who is responsible in the next hand-off. This milestone planning meeting is where we begin that process. Each of those people, the individuals that are in the meeting, are going to be asked:

- "What is the milestone you must complete in order for your specialty to hand-off the work to the next specialty or the end user?"
- "Can you and your team accomplish this milestone?" That's followed by a similar question,
- "Will you and your team accomplish this milestone for the individual you are handing off to?"

In the next set of meetings, we call these meetings "conversations", we begin working with the last planners, that is, those members of the above teams that are the individuals responsible for planning the actual work that is required to accomplish the milestone. Be sure to ask the leaders in this conversation for those people. We get into that in the next chapter.

Figure 2 - Showing one of the boards we are using in one of our construction projects. Note: the upper section is only milestones.

7 – Pull Planning

What?

Pull Planning is an intentional retrospective review of the final steps of the processes that lead to a conclusion. Pull planning is a decomposition process. We are imagining ourselves going through time backward from the point of the finished product or output.

All of us has performed a postmortem on a project that has gone poorly hoping to determine how we might have done better or, in very bad cases, how things went totally off track or even failed. We have had the dubious honor of taking the blame for poorly planning something that did not work, and the resources had to scramble to remediate the process, so the project would work. We have performed postmortems on project implementations that had to be backed out because they didn't work or worse yet, broke something else that was working.

One method we may have attempted to eliminate those results (particularly in very sensitive projects) is a premortem. This is a practice that starts by saying the project has failed, now come up with all the reasons you can think of that could have caused it to fail. (Portny, 2013) Not all retrospectives are from a negative perspective. However, all retrospectives will help us avoid negative outcomes.

We're spending a lot of time on this because we sometimes have to help the last planners understand what we are helping them do. In the last chapter we created the Conditions of Satisfaction. Those Conditions of Satisfaction describe what the last planners are working to deliver.

Healthcare Information Technology Integrated Project Delivery

We may want to begin by reminding them of the last go-live they participated in and ask questions like:

- What was the most challenging thing you overcame when the product went live? How could we (as a team) have done better?
- What was the last contribution you and your coworkers made to the product on the day before the go-live?
- What did you and your coworkers have to wait for from another team or vendor before you were able to perform the last task before the go-live?

We may also employ the five-why method. We are decomposing their past go-live. Once they begin to understand, we can employ the transference technique (Freud, 1895) but instead transfer the thought process from a project that has occurred in the past to a project that has not yet occurred. This will become easier the more frequently we and the other team members practice it.

We as Healthcare Information Technology Integrated Project Delivery Professionals are looking for the **hand-offs** between groups or teams. No team or individual does all the work in a project by themselves. No work is done in a vacuum. All project work performed in a vacuum fails. (Harvard Business Review, 2012) In our last chapter we identified specialty swim lanes, in this pull planning conversation we, as Healthcare Information Technology Integrated Project Delivery Professionals, are looking for the "links" where a product migrates from one specialty swim lane to another, we refer to this process as a **hand-off**. Hand-offs are critical to the success of the pull planning methodology. We begin by looking at the go-live and then looking backwards in time to what the last hand-off was prior to the go-live. The term go-live actually refers to the final hand-off where we, as information technology professionals, handoff the product to the end users. (LCI, 2017)

In the pull planning conversation our product is a network of commitments. Personal commitments are made between the party that is handing off and the party that is receiving the handoff.

Why?

Humans are notoriously poor planners from the rear. We have all heard the saying, "hindsight is 20-20". So why not start at the beginning looking backwards? By assuming the product is complete and decomposing the components and tasks that were required to complete the product, and by doing this collaboratively with all those parties that are associated with developing the components and performing the tasks we can better plan as a team.

It is important for us as Healthcare Information Technology Integrated Project Delivery Professionals to fully internalize the ideology of Lean Integrated Project Delivery. It is a very agile concept to include all the technical players of product development and composing those technical players into small teams that develop in sprints. The Last Planner System® is a cousin to the agile methodology but expands the idea of teams to include the end users and stakeholders in the planning. We do this because we are no longer project managers but facilitators of the team of individuals completing a project as a team. In larger projects more than one team or many teams may be working simultaneously to accomplish the overall project. The big room meetings that we have every two weeks or month calls in the leads or last planners of those teams. By forming this **team of teams** we can then work collaboratively to fine tune the team hand-offs so that the plan that was originated in the first meeting or conversation can continue to become more accurate and more cost-effective. This is the very heart of Lean Integrated Project Delivery. (Womack J. , 2003) (Ballard I. D., 2016)

In Healthcare Information Technology Integrated Project Delivery our goal and our mission is to facilitate a team or group of teams in their efforts to fine tune the design, methods, and plans of **their** project so that the least amount of work may be accomplished to efficiently and effectively deliver the project on time and within budget. Due to the iterative conversations among team members that we are facilitating, constant incremental performance and planning improvements are occurring within the team or teams. (McChrystal, 2015)

How?

Employing the tools above, and bringing as many of the last planners into his big conference room as we can find, push all the tables out of the way, move the chairs out of the way, and get everyone up on their feet.

Everyone participates in pull planning.

As the facilitator of this integrated project delivery, you should bring some things with you to this conversation. You should bring colored sticky notes, a different color for each specialty swim lane, and felt tip pens, enough for everyone. Distribute the colored sticky notes to each person who is contributing for their specialty. Each of the sticky notes is organized as shown in Figure 3 below. In the upper left corner of the sticky note are the initials of the individual who is responsible for the individual task that is listed in the center of the sticky note. At the bottom of the sticky note that individual should write a description of anything they need someone else to accomplish before they can perform the task that they have written in the center of the of the sticky note. In the upper right corner of the sticky note the duration in days of that task is written.

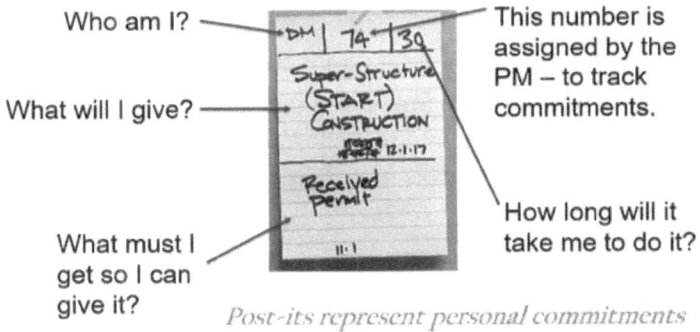

Post-its represent personal commitments

Figure 3 - How to format the sticky note

The sticky note represents an individual's commitment to fulfill an individual task in a duration of time. The section at the bottom of the sticky note which stipulates a requirement the individual has in order to accomplish the task that they have committed to should have a matching individual on a different sticky note who is committing to deliver that deliverable. As these individuals write the sticky notes they should stick them on the board in the order they initially think they are accomplished.

Remember, we are working backwards from the milestone, so the first sticky note on the board should be the **last task** performed before the go- live. The next sticky note on the board should be the task that enables or informs the last task prior to the go-live. Using this example, we should have a diamond on the far-right side of the board labeled "go live" and to its immediate left there should be a color sticky note which has an individual identifying the task that they are going to complete with the duration of that task, and that should be the last task before the go-live that enables or informs the go-live. And the next one to that enables or informs that task.

This exercise is repeated over and over as each of the individuals who are standing in the room as part of the specialties that are required to accomplish the product or service begin to decompose the tasks from the go-live to the inception.

As this process continues we begin to realize that we are building a network of commitments. It's not important at this part of the planning that the sequence of events is perfect. What is important is that all the events, or tasks, are identified. In this iteration of the discussion all the members of the team, even though they are from various specialties within and outside of the Healthcare Information Technology field, are talking and working together to identify all the tasks that must be completed to reach the milestone. As a Healthcare Information Technology Project Manager our role is coaching and encouraging these individuals to dig deeper into the planning concepts and identify the tasks that they not only need to accomplish but need to have accomplished so that they can accomplish their tasks.

When we feel this iteration of the conversation is beginning to wind down, it's time to interrupt the thought process of the individuals and start reviewing the sequence of the events.

Healthcare Information Technology In Motion

Healthcare Information Technology
Integrated Project Delivery

Again, we are the facilitators not the sequencers. The **team** now begins to sequence their own tasks. We are only facilitating them in making their own plan. No longer are we the task masters or task managers, but we are facilitating those individuals who are performing the tasks within the team so that they may identify the work that they must do and the work that the team must do in order to accomplish the task. Once the sequence has been set by the team, we validate the individual commitments that have been made on the sticky notes. We will have noticed during the process, individuals will have adjusted the durations that they believe it will take to get their jobs done because they are now working directly with other individuals for which they have been compensating unknowingly. Durations for individual tasks committed to by individuals during the pull planning process tend to get shorter with every iteration of pull planning. As the individuals become more comfortable with the technique and tools of the pull planning the process is **building trust** among the team.

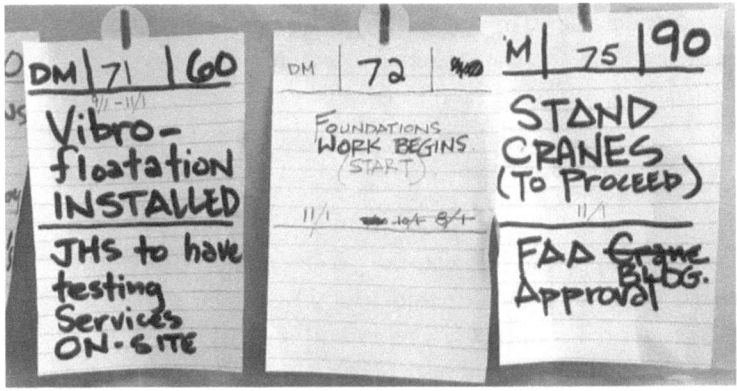

Figure 4 - Sequencing the sticky notes produces the plan

Healthcare Information Technology In Motion

55

Healthcare Information Technology
Integrated Project Delivery

As Healthcare Information Technology Integrated Project Delivery Professionals, we are no longer guessing what the durations were going to be or where the hand-offs are going to happen because we have determined both the hand-offs and the durations of those tasked with the entire team. We have facilitated the team in their ability to do their job effectively efficiently and in a timely manner. We have found through experience that early resistance quickly falls away as the team members begin to understand the value of the pull planning conversation.

Once this is complete, take the time to ask the participants if they have found any value in this conversation. One method for doing this is called a "plus / delta". This is a simple process in which standing at the board, we write a big plus on the left and over to the right we draw a triangle. Asking everyone in the room to participate, we ask them to identify things that we have done well in this meeting. We write that list under the plus. Then we ask everyone to identify things that we as a team could have done better, we write those items under the triangle. In this manner, the entire team participates in the lean practice of constant improvement as well as learning from each individual event as opposed to waiting until the end when we don't remember everything that happened during the project and it's therefore difficult to learn from it.

This concludes the pull planning conversation or meeting. However, it does not conclude our work as Healthcare Information Technology Integrated Project Delivery Professionals. It is our opportunity now that the team has identified all their tasks, committed to each other personally, estimated the durations of their tasks, and sequence the events - to create a commitment log. We found the easiest way to create a commitment log is to do so on a spreadsheet. There is no set rule of how the spreadsheet should be created however, the fields that are on the sticky notes as a minimum should be converted into columns on the spreadsheet. Using the current date and the durations provided by the individuals as committed to each other, it is relatively easy to produce a series of dates based on durations relative to the current date. By doing this, the Integrated Project Delivery Professional can quickly produce a timeline for the anticipated completion of the project.

Weekly, we should hold a 15-minute standup or teleconference meeting to review the commitment log. The purpose of this meeting is very agile in nature. We are not holding the meeting to beat people up, but rather to find out how the team can support each other in accomplishing their goals. As the facilitator we may find that an individual has stopped forward motion on their tasks because they have run into a constraint. It is our job as project facilitators to knock those constraints down so that the individuals may accomplish the tasks that they've committed to.

Pull planning is an iterative process. It is not a "once and done" exercise. If we can leave the sticky notes up on the board where we're working – we should do so, but if we can't it's worthwhile to find a manner of moving those notes in their correct order onto some type of media that allows us to move them from one point to another in case we need to have our next pull planning conversation meeting in another room. This can be done on roll paper that is readily available in places like Home Depot, or it can be done on poster board or can be done with a photograph from our cell phone and then move the sticky notes onto some paper that allows us to keep them in order. One manner by which the project manager can help themselves maintain the order by which they move the sticky notes from the board is to use the center part of the upper section of the sticky note to number the sticky notes for location sequence - see Figure 3 above.

The commitment log that we have created should either be posted in a commonly available place such as a shared network drive that all members of the project team have access to or e-mailed to all members of the project team. Figure 5 shows an example of a commitment log.

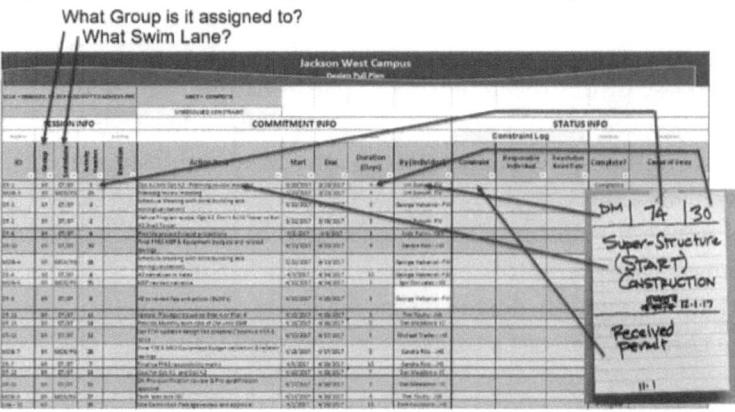

Figure 5 - A sample of a commitment log

When?

Pull planning conversations should begin as soon as it is possible following the milestone planning conversation. It is often difficult, particularly at the beginning, of altering the manner of project management to Lean Integrated Project Delivery to get everyone who needs to be in the meeting to actually commit to coming to the meeting. In Healthcare

Healthcare Information Technology
Integrated Project Delivery

Information Technology we all tend to be very busy. We as Healthcare Information Technology Integrated Project Delivery Professionals need to be sensitive to this demand on everyone's time. However, we should plan on 2 to 4 hours for the initial pull planning conversation. It is important that the team members understand and therefore it is important that we as Integrated Project Delivery Professionals explain that this intense commitment of time and people actually saves the information technology organization more time and more effort than the conventional method of having many small meetings over many days to accomplish the same goals. By the end of the pull planning conversation meeting they will have understood that not only did we save the organization time and effort in the planning but also that the planning accomplished was much more effective and more accurate than conventional project management planning.

Figure 6 - Can't find a room? - Use the hall!

8 – Technical Design Planning

What?

The Last Planner System® calls this step Make Ready Planning. The thought being, after the Pull Plan conversation the last planners would then return to their respective specialties and plan to get the work done. In construction that often means getting all the materials ordered and delivered so the work can actually begin.

Similarly, in Healthcare Information Technology there is a need for those who have just committed to getting a task (or set of tasks) complete to produce a detailed design (plan) for how they are going to produce, build, or implement their commitment. In the same method as Agile Project Management we are helping these technical specialists "chunk down" the tasks into smaller deliverable "sprints". Although the Agile term "Sprint" refers to a team coding effort of delivering working software, a similar work flow methodology applies here. Although the team may be doing anything from determining the network infrastructure changes and additions needed to actually developing code to meet the Conditions of Satisfaction, most often the work effort is implementing an upgrade, enhancement, or new module of a vendor supplied system such as Cerner or Epic. Even if it is a sub-system associated with a specialty field like Lab or Pharmacy there is still the need to customize and integrate with the larger system.

In this phase the detailed specifications of the desired project deliverable are gathered and the individual performers who will be performing the work are determining if they have everything they need to accomplish the deliverable task output as well as when they will be able to schedule it. User group meetings can also be helpful in this phase to determine the specific needs of the delivered product. (Ballard I. D., 2016)

In one environment we may a have a spreadsheet that we refer to as "The Room and Bed Master" this is a device that is used by the individuals in Information Technology to gather the requirements from the end users to assure that each room is allocated with each of the electronic medical record tools as well as cost centers, billing codes, ICD 10, and other specifications that are needed by the electronic medical record to accurately and efficiently provide the caregivers the information they need relative to the specific activity that will take place in room. This is but one tool of many and it is used here as an example of things that may be necessary to **make work ready**.

As Healthcare Information Technology Integrated Project Delivery Professionals, we are employing management by walking about in this phase. We are not actually doing the work. We do need to assure that any constraints that are encountered by the team are knocked down. Particularly at the beginning of this phase as the individuals who are part of the pull planning session are learning how to implement the commitments they made in the pull planning session. Our roll now should be facilitating their understanding and learning.

There is a fine line between facilitating learning and meddling in their affairs. We as Healthcare Information Technology Integrated Project Delivery Professionals must be aware that this line exists and exert every effort to not cross that line. We have asked these team members to make commitments to each other not to us, therefore, our purpose in coaching these team members must be clearly understood by both them and us that we are facilitating not micromanaging.

Why?

Everything associated with Milestone Planning and Pull Planning answers the question – "What SHOULD we do?"

In this phase we are answering the question – "What CAN we do?"

Similar to agile project methodology our adaptation of The Last Planner System of Production Control® has reduced the deliverable milestones to individual work unit deliverables. This enables teams of information technology professionals to work in concert with each other and other workloads to accomplish individual tasks that contribute to the greater project outcome. Until we implemented this methodology, Healthcare Information Technology Project Managers were relegated to the wasted efforts of attempting to understand the professional needs of every discipline and specialty that Healthcare Information Technology supports in the clinical environment. This waste was compounded by the time and effort consumed in the back and forth discovery that the Project Manager was performing to find what work needed to be performed by whom and when. Even as I write the above sentence I realize how totally unattainable that objective is. There is no way that any Healthcare Information Technology Project Management Professional can assimilate, learn, and understand all the facets of all the professionals that are supporting all the specialties and all the professionals in the clinical arena of a major hospital. Even the most advanced medical doctors have given up trying to be all things to all people. We as Healthcare Information Technology Integrated Project Delivery Professionals must also realize that we cannot be all things to all people. However, we can coach teams of people who, together can more nearly assimilate the needs of the entire healthcare organization.

Through Healthcare Information Technology Integrated Project Delivery, we are more nearly able to bring planning into an accurate assessment of the workflow and effort that will be required to accomplish any given goal. This **make ready planning** or as we're calling it in our adaptation of this methodology the Technical Design Planning, is the phase where we begin to learn what each of those individuals actually does in their cubicle to make the entire Information Technology team work.

As this phase is being implemented, you may find it is worthwhile to begin a constraint log identifying those constraints encountered during this make ready planning. This constraint log is an important learning tool as we plan for the future because we will run into similar constraints in similar work efforts in future projects. Identifying and logging these constraints will enable us to better pull plan in future projects. Additionally, constraints serve as clues to unseen linkages, or overlooked hand-offs, between the projects that we are working on and other projects that other project managers may also be working on.

The intention of this system of project management and the fundamentals of the practices involved are to produce predictable uninterrupted workflow. By creating a comprehensible set of commitments that connect the work of the specialists to the promises of the project to the client, this methodology coordinates interactions toward success. (Seed, 2017)

Healthcare Information Technology In Motion

Healthcare Information Technology
Integrated Project Delivery

How?

Referring back to the Conditions of Satisfaction we will work intentionally with the last planners to help determine what work is needed to accomplish the goal and produce the committed outcome in the duration of time committed to. The most difficult challenge that any technical person ever encounters is we tend to "over think" things. **Thinking is what we do for a living**. Much planning is done in the technical person's own mind. Therefore, this is a challenge to get those technical people to work in a team format and **groupthink**.

The Japanese term for this is Gemba. Gemba, means "the real place" that is generally used to refer to the place **where value is created**. In six Sigma managers are taught to do a "Gemba Walk" which refers to walking about where the work is actually getting done. In Lean processes team areas where technical staff can work together with little or no separation are referred to as a Gemba. In Latin it would be **"locus opus"**.

Lean practitioners through many years have determined that grouping people together by the work they're trying to accomplish, not necessarily by their specialty or discipline, is the most effective way for a team to accomplish an individual task or group of tasks to produce an output product. (Womack J. , 2003)

We as project managers have struggled with this throughout our careers; it will be no different now, but now we have the tools to create the learning experience that will help us overcome this hurdle.

Healthcare Information Technology Integrated Project Delivery

Through the iterations of our conversations in this Last Planner® methodology we have experienced improvements in the commitments that are made, timelines that are predicted, cost of producing the outcome, and the quality of the output product. We don't expect our first iterative attempt at this methodology to be perfect, but we do expect to get better with every iteration always moving toward a more perfect outcome.

We coaching and facilitating to help the doing be more efficient.

An important element of Make Work Ready phase of the Last Planner® methodology is discovering what needs to be ordered, when it needs to be ordered so that lead times do not impact deliverables, and how to pay for orders so the project completion is not affected.

When?

Technical Design Planning or Make Ready Planning as it is called The Last Planner System®, is a daily, weekly, or biweekly formal event completed by the last planner and the technical team associated with that individual. Everything about Lean Integrated Project Delivery is associated with team building, team commitment, team performance, and team delivery. Each individual is working toward their best performance based on their commitment to the team and the commitments from the other members of the team that they too are working toward their best performance. Always working to learn and be better.

Semper Melius!

Healthcare Information Technology In Motion

9 – Weekly Work Planning

What?

If there is a nitty-gritty to go with the nuts and bolts, the weekly work planning section is the nitty-gritty. This is where we find out how reliable our promises and commitments have been to each other. This does not imply that commitments made were made falsely as much as it says that we are not necessarily good at estimating when we start out. Each iteration of pull planning and make work ready planning fine-tunes the commitments and timelines and therefore the expectations of the project. Healthcare Information Technology projects traditionally encompass multiple disciplines and specialties, each typically comes to the project with independent and mutually exclusive goals. These differences can lead to misunderstandings that in turn lead to incorrect work being done and therefore multiple reworks within the project. Each of us as Healthcare Information Technology Integrated Project Delivery Professionals has been confronted with the need to back out a go-live because it either didn't work or it broke something else that was already working.

Lean Integrated Project Delivery empowers performers to offer conditional promises or commitments which leads to negotiation. We are very familiar with these conditional promises in Information Technology. In coding we have a conditional branch we refer to as "if, then, else". This, in its own way defines conditional promising. In his book "Conversations For Action", Fernando Flores modeled basic action workflow as a series of mutual promises between customers and performers. (Flores, 2012) In his model of the **basic action workflow loop**, each individual workflow is a structure of commitments that constitutes transactional agreements between two parties within the loop. What we have created through the previous chapters of planning is a practical "network of commitments". If we stand back from the wall where we have performed the pull plan what we see is a "project network diagram"! (Project Management Institute, 2013)

Healthcare Information Technology projects tend to be a complex matrix of systems that work together, interact, and have multiple integrated budgets. Many decisions are made when developing and delivering projects to meet the Conditions of Satisfaction. No one Healthcare Information Technology professional, nor even a small group, can make all the necessary decisions to produce a successful project outcome. We are confronted with too many issues, technical and medical skills, inputs, software knowledge, options, and regulations. A Lean Integrated Project Delivery project uses Cluster Groups to better manage this task.

Cluster Groups are small groups of individuals assigned to a related task, usually grouped by specialty. These are similar to the agile **scrum team**. The daily and weekly work units are performed in these Cluster Groups. (Since we are Healthcare Information Technology Integrated Project Delivery Professionals, feel free to use the term scrum team if you're uncomfortable with the term Cluster Group.)

Cluster Groups should be formed around whatever grouping is appropriate to the project. The cluster group should collaboratively form to define objectives and innovations that it will bring to the project.

Healthcare Information Technology In Motion

Healthcare Information Technology Integrated Project Delivery

As with the agile scrum team, daily huddles may prove to be of great value as a tool for the cluster team to plan their daily work and assure the project schedules, individual commitments, and output products are completed on time. Daily huddles are not necessarily problem-solving or planning meetings. Rather, they are the means for the team to surface unresolved issues collaboratively. These should be standing meetings, as in *standing up*, to force the meetings to stay short and agile. (Agile Alliance, 2017)

The objective of this phase is to focus on effective, efficient, quality work. This daily huddle helps to maintain workflow progress, identify constraints, measure performance, and learn from variances in the plan. (Seed, 2017)

Why?

Weekly check-in calls, of the duration no longer than fifteen minutes, along with daily huddles at the Gemba or **Locus Opus,** are the root of this cycle of Lean Integrated Project Delivery. If you've been in Information Technology for any time at all, you are aware of the OSI seven-layer model. The Last Planner System of Production Control® is similarly constructed. At this layer the work effort is referred to as **"will".**

The question that is answered at this layer, in this conversation, is:

"What will we do?"

How?

There should be a limited scope of work effort being planned in this phase. The individuals who are now preparing for their task delivery should be looking no more than two weeks in advance.

The two questions that are being asked and answered in this phase are:

1. "What do I need to do to accomplish my deliverable task in the next 1 to 2 weeks?"

2. "Is there any obvious constraint that needs to be overcome or eliminated in order to accomplish my deliverable task in the next 1 to 2 weeks?"

The answers to these two questions sum up the list of things to do for the individual performers over the short duration look ahead period.

This phase or layer follows the Make Work Ready phase or layer, in which, the questions was answered – "What **can** we do?" Therefore, in the previous layer or phase the team has already broken the workflow down into digestible bites, or chucks so that the Cluster Groups can efficiently complete the work. These chunks should be no longer than two weeks duration.

Learn and practice the performance of Percent Plan Complete this helps gauge the reliability of the planning system as we apply it. Percent Project Complete is the number of planned activities that we have completed divided by the total number of planned activities. The outcome is expressed as a percentage.

Percent Project Complete measures the extent to which the last planners (line managers and supervisors) commitment (in the WILL segment) was realized (Ballard H. G., 2000). Project Percent Complete tells us if the planning process is reliably predicting what will actually be completed. This tool's output can be misleading if the tasks completed exceed a duration of two weeks.

When?

Weekly, daily.

Weekly: As with a scrum team the Cluster Group is working together to affect a change or build a product. A typical week would be similarly modeled.

Monday – Morning check-in session (15 minutes) formally called, stand-up, and on the calendar.

Wednesday – Morning check-in session (15 minutes) formally called, stand-up, and on the calendar.

Wednesday – Afternoon (as needed, minimum every two weeks) pull-planning session against the plans from the Big Room. Review of the Constraint Log, review variances. Perform Project Percent Complete.

Friday - Morning check-in session (15 minutes) formally called, stand-up, and on the calendar.

Daily applying the Deming PDSA methodology of systematic learning and continually improving our products, processes, and service. (Deming W. E., 2017) This is accomplished in the form of daily huddles. These huddles should take place as close to the location where the work is actually happening as possible. Lean practitioners often refer to the work location as the "Gemba", and again the Latin is "Locus Opus".

Figure 7 The Deming Wheel

A retrospective learning event should occur with the completion of every deliverable or monthly to learn how we could "do this better".

Go check out his video:

www.youtube.com/watch?time_continue=18&v=_BCPdIHsPog

10 – Learning & Improving

What?

People are more important than the processes we use to accomplish the tasks or projects people want to accomplish. Individuals are more important that the tools we develop. Technology was built for people. People were not invented to make the technology work.

Completed projects are more important than the documentation of the project as it is being done. The **effect** (consequence) of Information Technology is more important than the **affect** (influence) of the project processes of technology as it is being applied to gathering information. The **result** of being able to develop knowledge from the information derived from the data is more important than reporting about how we got there.

Collaboration with the people who work for the patients to provide healthcare is more important than the management reports about what those people are doing to provide such healthcare. The knowledge provided by the Healthcare Information Technology will provide the science to end disease as we know it and benefit all humankind. Reporting is and continues to be an important learning tool, teamwork is the most effective method for improvement.

As opposed to inventing a plan that looks forward we understand there is a need to be flexible and **embrace change**. We plan more expertly by working retrospectively from the vision's "live perspective" than from the starting point. We plan in more detail, and better detail, as we get closer to the objective. We expect the plans to change and we expect the path to alter as we move toward a more perfect

Healthcare Information Technology Integrated Project Delivery

application of the Information Technology that will better serve the healthcare professionals who are in turn building better health outcomes for the population.

This follows closely the ideology of the Agile Manifesto. (Jeff Sutherland, 2017)

Knowledge (not Information Technology) is power. The power we seek is the power to heal.

We recognize that we will never know less about the processes and steps to accomplish an outcome than at the onset of the project. (Layton, 2012)

Learning as a result of studying the processes and improving on the work we are doing is core to the system of Integrated Project Delivery. We make every effort to learn from every success and every failure at every step along the way of the project.

Always learning - semper in doctrina.

Learning allows us to take informed actions to **improve** our plans and reduce wasted time and effort in accomplishing our goals. (Deming W. E., 2017)

Why?

Our highest priority is to satisfy the Conditions of Satisfaction as early in the process and as efficiently as is possible. We expect changing requirements as we proceed toward the Conditions of Satisfaction because the Healthcare Industry is always changing and improving. As we learn from and evaluate the work we are doing we are better able to produce elegant outcomes. Elegant outcomes are models of simplicity accented by maximization of the work not done.

Healthcare Information Technology In Motion

Healthcare Information Technology
Integrated Project Delivery

Our objective it to eliminated wasted effort and activity in the process of accomplishing our goals. (Agile Alliance, 2017)

> "There are four purposes for improvement: Easier, Better, Faster, and Cheaper. These four goals appear in the order of priority."
> Shigeo Shingo

We strive to get better and be better at our profession every day, in every iteration of our effort we apply what we have learned and become more efficient and effective.

Getting better - Questus melius.

We are always growing professionally, and increasing our competence. We are always getting better.

Always better - Semper melius.

How?

Weekly Check-In Calls:

- Is your commitment complete?
- If not when will it be complete?
- Is there a roadblock that is keeping it from being complete?

15 Minute Phone Call

This is not about what you did wrong! It's about how the team can help get it done!

This is driven from the commitment log.

Healthcare Information Technology
Integrated Project Delivery

Bi-Weekly (or Monthly) Big Room Meetings:

- Bring the pull-plan back and revisit the activities.
- Mark off the sticky notes that have been completed to date.
- Pull plan new milestones and discoveries.
- Revisit schedules and commitments.
- Apply the Percent Project Complete.
- Perform the plus – delta exercise.

When?

Always. Every step of the way, every day, every week, every meeting, and always. Lean Integrated Project Delivery affects all aspects of our career.

11 – Wrap Up

What?

Congratulations! You made it to the last chapter. We thought we would use this chapter to close the project, but we aren't closing a project we have "task mastered" or project managed to its end. No, we have been coaching a team of professionals in a collaborative, integrated, and interactive methodology and … like a sports team … the team has delivered an integrated project. Well done!

We left some stuff out of this book that you will need to study for yourself. I left these subjects out of the book because they are not being adapted specifically to Healthcare Information Technology Integrated Project Delivery. From the beginning of this book we made the assumption we were all Professional Project Managers. We assumed the reason for reading this book was because the current methods we were using were not working.

There is much more we can do though as Lean Integrated Project Delivery agents of change. This is a career change that we are making as a team. We must go "all in" and commit ourselves to constant and never-ending improvement. This we must apply to our very lives. We don't need to expect everyone to suddenly understand and support us, but we are obligated, now that we have seen the value, to keep constant forward thrust toward the goal of Lean Enterprise Processes in Healthcare Information Technology. What we do is too important to the human race for us to leave it on the side and continue down the path that obviously doesn't work.

Healthcare Information Technology Integrated Project Delivery

We, however, are still obligated to report status to those who are observing our efforts. There is another Toyota Production System model that we'd like to introduce. This is known as the "A3". The A3 is so named because it is on A3 size paper – 11" X 17" (US). John Shook's book "Managing to Learn" is a good read, it breaks it down and illustrates its use. One can also get sufficient information by performing a quick Internet search of "A3 Problem Solving Process". To summarize the A3 breaks down a report on a single page that is formatted to cover the Deming PDSA Wheel. Reporting is recommended to be **visual analytics** as opposed to words and numbers. Toyota posts these on the wall or bulletin board. The idea is not that different that the "grease board" of days gone by. Of course, we now use electronic status boards in their place. As a Healthcare Information Technology Integrated Project Delivery Professional, we would like to see these electronic A3s be used as KANBAN boards showing the status updates of the work our teams are performing. (Shook, 2008)

The A3 also captures the budget and cost performance visual representations. While I have been writing this book the Project Management Institute has published PMBOK Version 6. My copy is in the mail, but I did download my free member's copy and reviewed it. If you are in an environment where your work includes monitoring the spend of the Information Technology Division, or if you have a project that is governed by a Purchase Order limitation or other monetary limitation, the PMI has published a number of mathematical formula to assist in calculating the financials. Most Healthcare Information Technology Project Managers I have talked to don't use those calculations because the work they do tends to be contract driven to software providers like

Healthcare Information Technology Integrated Project Delivery

Cerner or Epic. Where this does apply is generally in the areas of infrastructure and data center, where I have found a simple burn-down chart makes a good visual representation of the situation. Burn-down charts work well in the A3 methodology of reporting.

Why?

Your Healthcare organization may just jump on board, don't rule it out! It's very possible. Once the teams begin to form and the efficiencies begin to happen, people will begin to demand that we perform our projects in an Integrated Project Delivery methodology because we all start to realize the "once and done" benefits of incorporating deep collaboration with all the affected parties.

This method will also help integrate Information Technology into the systems of Healthcare from the perspective of the providers and clinicians. We all begin to feel like we are on the team together to accomplish the goals.

How?

The method of collaboration and planning explained in this book assures that the parties are working together and staying informed. The parties are making personal commitments to each other as opposed to waiting for another department to do their part for the project. People are working together with each other's schedules and workloads to accomplish the needed elements of the project. The team owns the project, the team owns the commitments, the team owns the budget, the team DELIVERS the project to the Healthcare System.

When?

Let's get started today. Get a copy of this book and give it to your CIO, CTO, or COO. This is a proven methodology,

Healthcare Information Technology
Integrated Project Delivery

proven globally in construction and we are proving it in one of the largest Public Healthcare Systems in the United States.

Healthcare Information Technology In Motion

12 – Visually Speaking

Figure 8 - The Five Phases of The Last Planner System

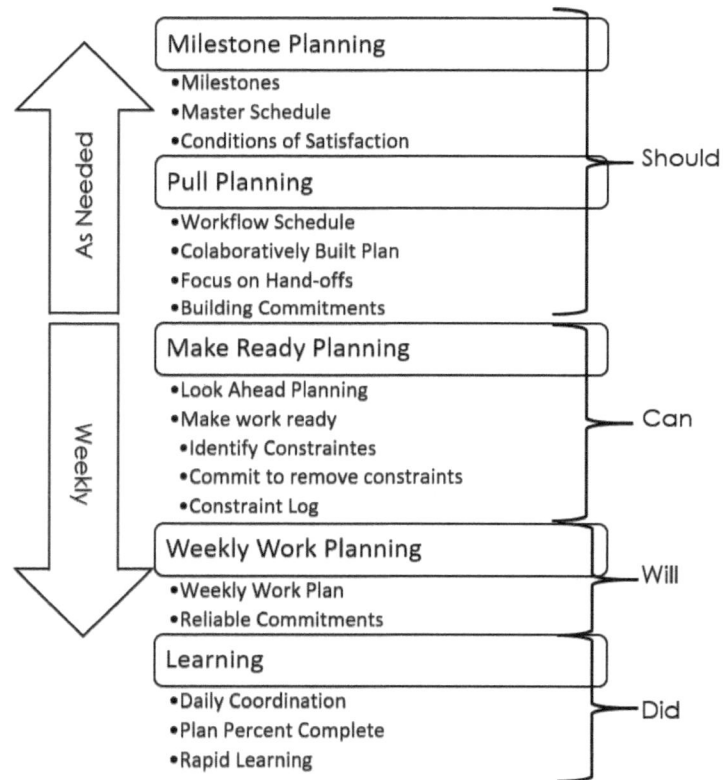

Healthcare Information Technology
Integrated Project Delivery

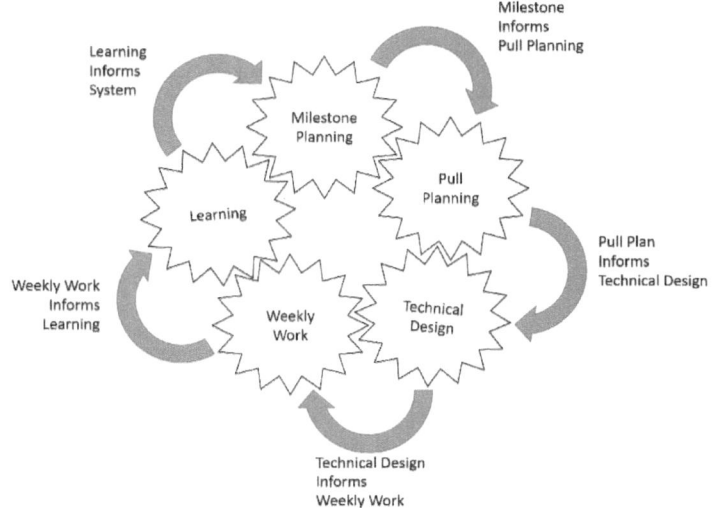

Figure 9 - The Learning Loop

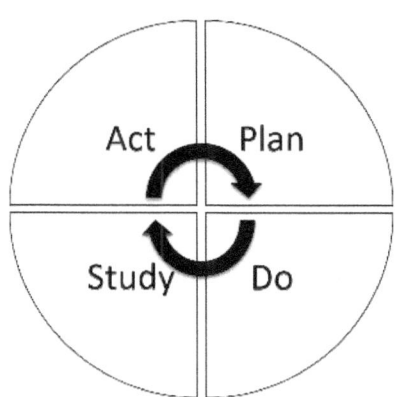

Figure 10 - The Deming Wheel

Healthcare Information Technology In Motion

Healthcare Information Technology
Integrated Project Delivery

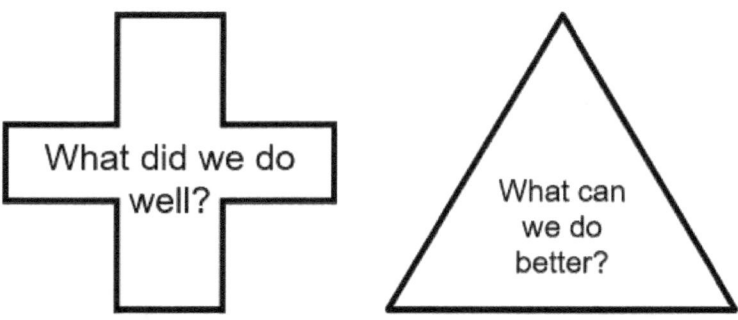

Figure 11 - The Plus - Delta Questions

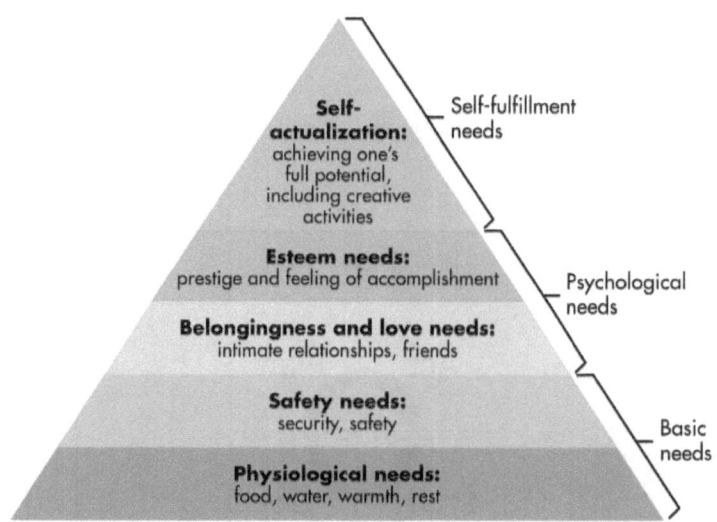

Figure 12 - Maslow's Hierarchy of Human Motivation

Healthcare Information Technology
Integrated Project Delivery

Figure 13 - The Flores Loop

Healthcare Information Technology In Motion

Appendices

Manifesto for Agile Software Development

We are uncovering better ways of developing software by doing it and helping others do it.

Through this work we have come to value:

- **Individuals and interactions** over processes and tools
- **Working software** over comprehensive documentation
- **Customer collaboration** over contract negotiation
- **Responding to change** over following a plan

That is, while there is value in the items on the right, we value the items on the left more.

Copyright • Kent Beck • Mike Beedle • Arie van Bennekum • Alistair Cockburn • Ward Cunningham • Martin Fowler • James Grenning • Jim Highsmith • Andrew Hunt • Ron Jeffries • Jon Kern • Brian Marick • Robert C. Martin • Steve Mellor • Ken Schwaber • Jeff Sutherland • Dave Thomas •

Healthcare Information Technology
Integrated Project Delivery

The Slide Presentation used at JHS

Integrated Project Delivery

Applied Last Planner® Methodology
In a Healthcare IT environment

Jackson HEALTH SYSTEM   PUBLIC HEALTH TRUST

Miracles made daily

Key Concepts

1. Traditional planning methods are unable to produce predictable workflow

2. Workflow reliability directly affects Health System project delivery and cost

3. All plans are forecasts, all forecasts are wrong, further in advance = more wrong, more detail = more wrong

4. Collaborate – Really collaborate!

Miracles made daily   Jackson HEALTH SYSTEM

Healthcare Information Technology In Motion

Last Planner® Principles

- Plan in greater detail as you get closer to doing the work.
- Produce plans *collaboratively* with those who will do the work.
- Reveal and remove constraints on planned tasks as a team.
- Make and secure reliable promises.
- Learn from breakdowns and failures.
- Structure the work to achieve smooth workflow.

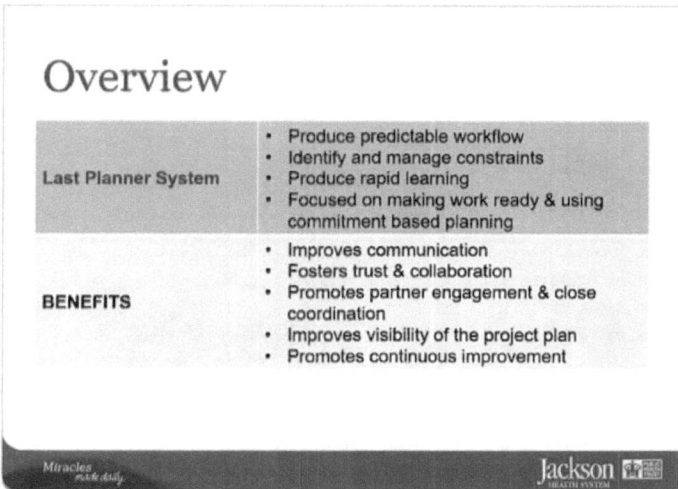

PM - The job of project control is to determine whether or not the project is achieving its objective.

Manager - The job of production control is to *do* what is needed to achieve project objectives.

Miracles *made daily*

Jackson HEALTH SYSTEM

Overview

Last Planner System	• Produce predictable workflow • Identify and manage constraints • Produce rapid learning • Focused on making work ready & using commitment based planning
BENEFITS	• Improves communication • Fosters trust & collaboration • Promotes partner engagement & close coordination • Improves visibility of the project plan • Promotes continuous improvement

Miracles *made daily*

Jackson HEALTH SYSTEM

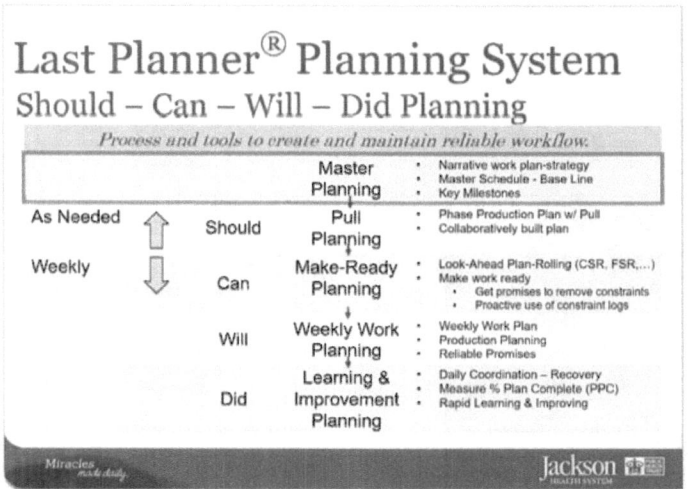

Last Planner® Planning System
Should – Can – Will – Did Planning

Process and tools to create and maintain reliable workflow.

		Master Planning	• Narrative work plan-strategy • Master Schedule - Base Line • Key Milestones
As Needed ⇧	Should	Pull Planning	• Phase Production Plan w/ Pull • Collaboratively built plan

Weekly ⇩	Can	Make-Ready Planning	• Look-Ahead Plan-Rolling (CSR, FSR,...)
			• Make work ready
			  • Get promises to remove constraints
			  • Proactive use of constraint logs

	Will	Weekly Work Planning	• Weekly Work Plan • Production Planning • Reliable Promises
	Did	Learning & Improvement Planning	• Daily Coordination – Recovery • Measure % Plan Complete (PPC) • Rapid Learning & Improving

Miracles *made daily*

Jackson HEALTH SYSTEM

Develop a Master Schedule

1. Identify milestones important to the stakeholders – especially immovable dates.
2. Gauge the feasibility of completing the work in the allotted time.
3. Establish the sequential relationships between activities.
4. Test the overall project execution strategy

Miracles *made daily*

Jackson HEALTH SYSTEM

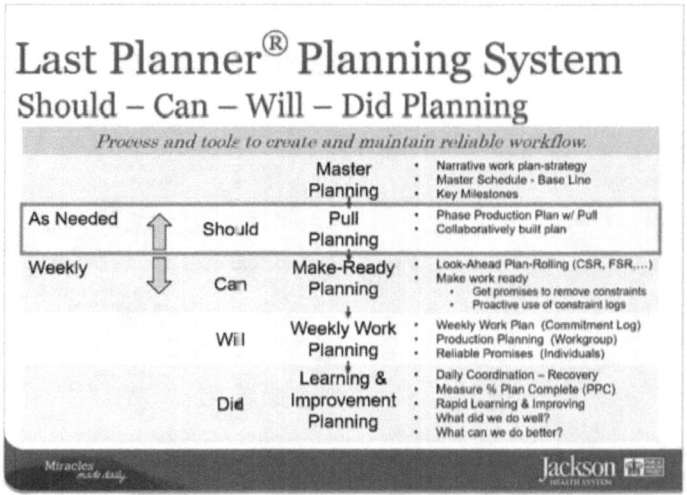

Last Planner® Planning System
Should – Can – Will – Did Planning

Process and tools to create and maintain reliable workflow.

		Master Planning	• Narrative work plan-strategy • Master Schedule - Base Line • Key Milestones
As Needed ↑	Should	Pull Planning	• Phase Production Plan w/ Pull • Collaboratively built plan
Weekly ↓	Can	Make-Ready Planning	• Look-Ahead Plan-Rolling (CSR, FSR,…) • Make work ready 　• Get promises to remove constraints 　• Proactive use of constraint logs
	Will	Weekly Work Planning	• Weekly Work Plan (Commitment Log) • Production Planning (Workgroup) • Reliable Promises (Individuals)
	Did	Learning & Improvement Planning	• Daily Coordination – Recovery • Measure % Plan Complete (PPC) • Rapid Learning & Improving • What did we do well? • What can we do better?

Miracles *made daily*

Jackson HEALTH SYSTEM

Pull Planning Meeting Tips

- All parties in a "Big Room"
- Start with "target milestone" on the far right.
- Add post-its when requested by analysts, specialists.
- Add post-its requested by the manager.

One by one, add tasks to the left. Work from right to left – using "pull mechanism". Each post-it asks questions –

1. "To start this task, I must have XYZ task done."
2. "What one completed task allows me to start?"

Miracles *made daily*

Jackson HEALTH SYSTEM

Pull Planning Detail Tips

It's **NOT** about every task you will complete.	It **IS** about defining hand-offs.

- Map enough of your workflow for others to follow.
- No task duration longer than the Look Ahead Duration. (CSR, FSR ... etc.)
- Tasks specific enough to communicate clearly, verify completion, and conditions of satisfaction.

Miracles *made daily.* Jackson

Pull Planning post-its   #1

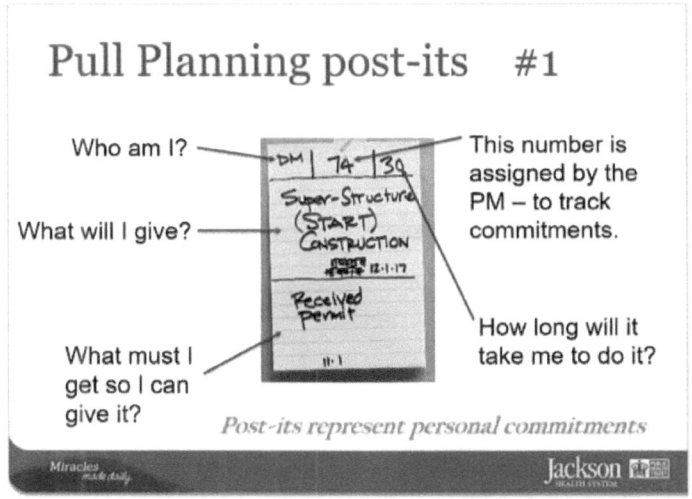

Who am I?

This number is assigned by the PM – to track commitments.

What will I give?

What must I get so I can give it?

How long will it take me to do it?

Post-its represent personal commitments

Miracles *made daily.* Jackson

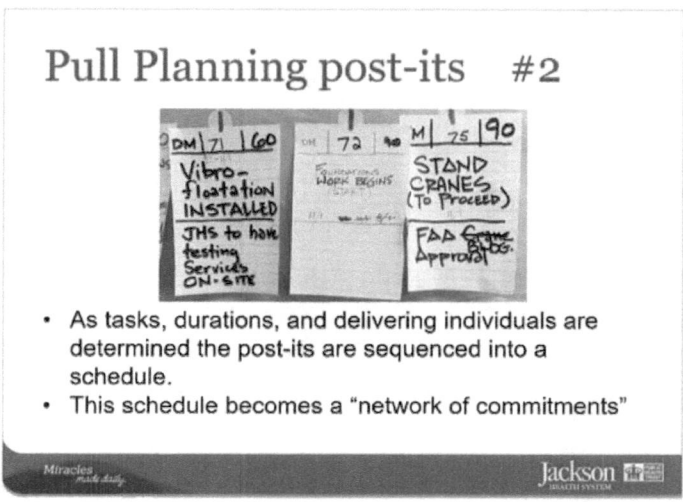

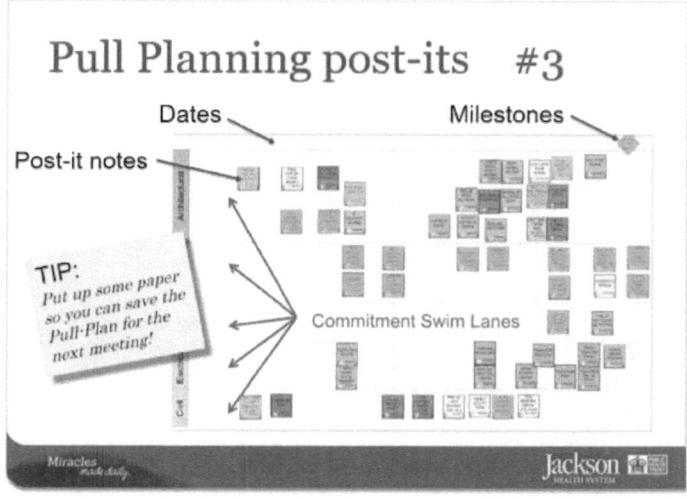

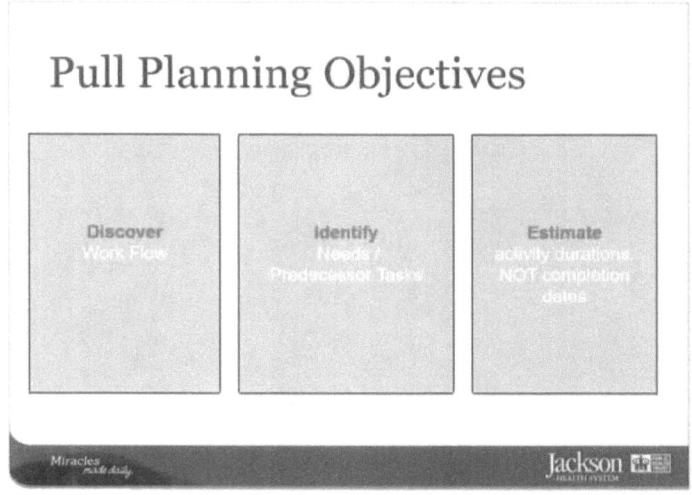

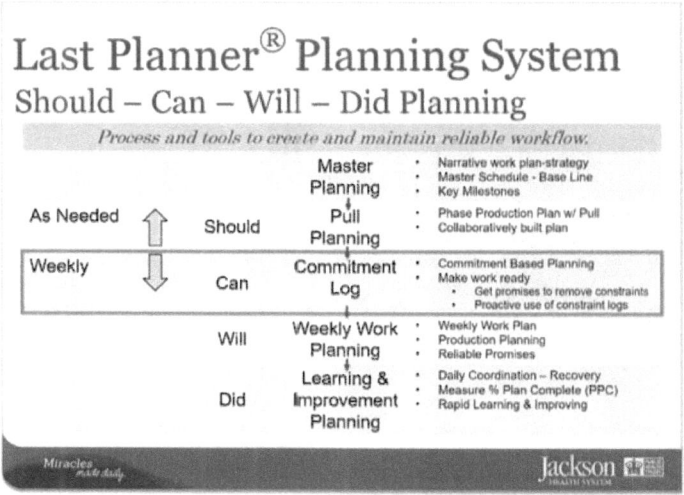

Healthcare Information Technology In Motion

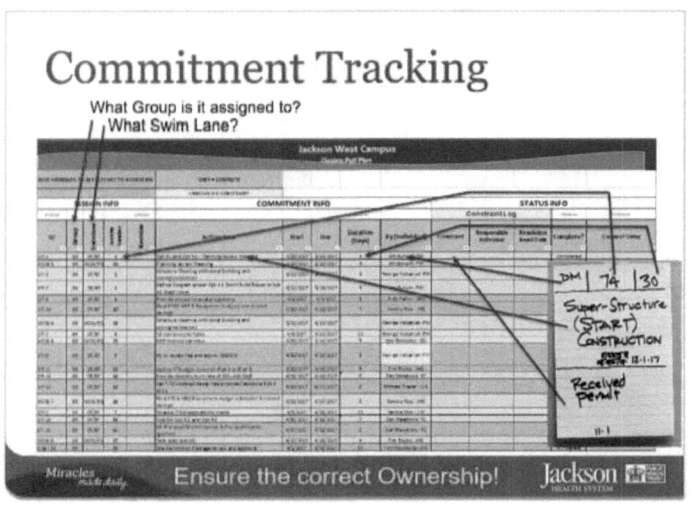

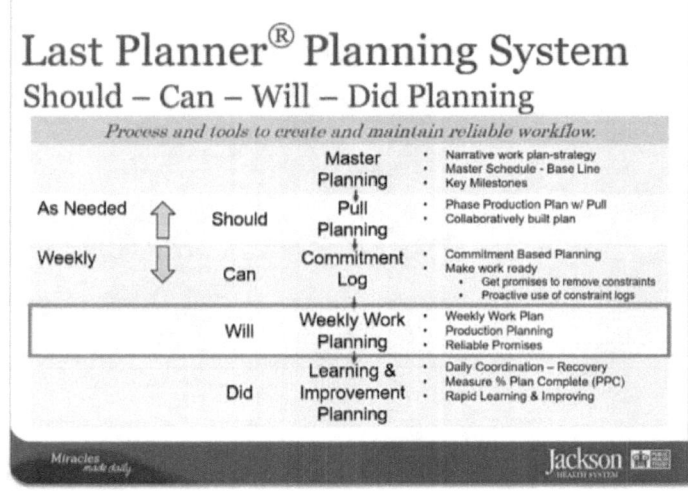

Weekly Check-In Call

15 Minute Phone Call

- Is your commitment complete?
- If not when will it be complete?
- Is there a roadblock that is keeping it from being complete?

This is not about what you did wrong!
It's about how the team can help get it done!

Driven from the commitment log

Miracles made daily

Jackson HEALTH SYSTEM

Last Planner® Planning System
Should – Can – Will – Did Planning

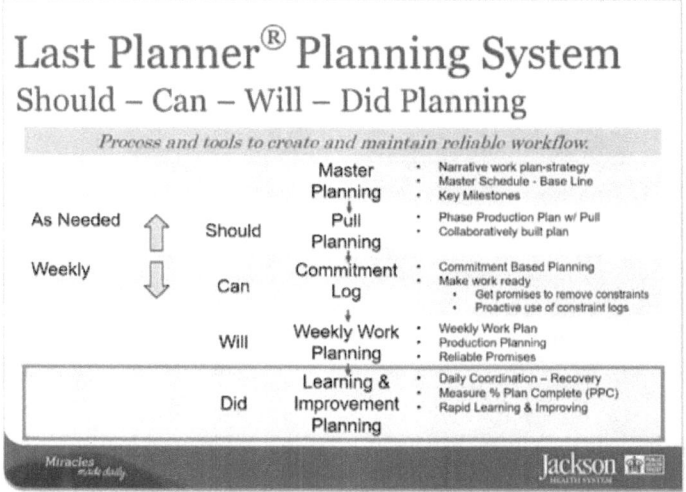

Process and tools to create and maintain reliable workflow.

As Needed ⬆	Should	Master Planning	• Narrative work plan-strategy • Master Schedule - Base Line • Key Milestones
		Pull Planning	• Phase Production Plan w/ Pull • Collaboratively built plan

Weekly ⬇	Can	Commitment Log	• Commitment Based Planning
			• Make work ready
			  • Get promises to remove constraints
			  • Proactive use of constraint logs

	Will	Weekly Work Planning	• Weekly Work Plan • Production Planning • Reliable Promises
	Did	Learning & Improvement Planning	• Daily Coordination – Recovery • Measure % Plan Complete (PPC) • Rapid Learning & Improving

Miracles made daily

Jackson HEALTH SYSTEM

It's all about constant improvement!

Bi-Weekly (or Monthly) Meeting Tips

- Bring the Pull-Plan back.
- Mark off what has been completed to date.
- Pull-Plan new milestones or discoveries
- Revisit Schedules and Commitments
- Calculate planning effectiveness
  (Number of commitments complete / commitments made)

Miracles *made daily*

Jackson HEALTH SYSTEM

Bi-Weekly (or Monthly) Meeting Tips

Typical Bi-Weekly Meeting Agenda -

1. Ice breaker – 15 Minutes
2. Hot topics – things happening right now! – 30 Minutes
3. Review Commitment Log – 15 Minutes
4. Coding and customizations discussions – 90 Minutes
5. Six weeks look ahead (Pull Planning) – 30 Minutes
6. New Business – 10 Minutes
7. Next meeting – agenda – 10 Minutes
8. Plus / Delta – 10 Minutes
9. Adjourn.

Miracles *made daily*

Jackson HEALTH SYSTEM

Healthcare Information Technology In Motion

Healthcare Information Technology
Integrated Project Delivery

Healthcare Information Technology In Motion

References

Agile Alliance. (2017). *https://www.agilealliance.org/agile101/the-agile-manifesto/*. Retrieved from Agile Alliance: www.agilealliance.org

Ammer, C. (2003). *The American Heritage Dictionary of Idioms.* Houghton Mifflin Company.

Ballard, H. G. (2000). *The Last Planner System of Production Contrl.* Birmingham, UK: School of Civil Engineering - Faculty of Engineering - The University of Birmingham.

Ballard, I. D. (2016). *Target Value Design - Introduction, Framework, and Current Benchmark.* Arlington, VA: Lean Construction Institute.

Cerner. (2017). *Cerner.* Retrieved from Cerner: www.cerner.com

Davidson, D. (June 22, 2015). *Last Planner System Business Process Standard and Guidelines.* Phoenix, AZ: LCI Israel.

Deming, E. (1994, Nov 19). *The New Economics.* Cambridge, MA: MIT Center For Advanced Educational Services. Retrieved from The Deming Institute: https://deming.org/explore/so-pk

Deming, W. E. (2017, November 18). *PDSA Cycle.* Retrieved from The Deminigs Institute: https://deming.org/explore/p-d-s-a

Dodson, C. L. (1865). *Alice's Adventures in Wonderland.* London: Macmillan & Co.

Flores, F. (2012). *Conversations For Action and Collected Essays.* North Charleston, SC: Createspace Independent Publishing.

Fowler, S. (2014, Nov 26). *What Maslow's Hierarchy Won't Tell You About Motivation.* Retrieved from Harvard Business Review: https://hbr.org/2014/11/what-maslows-hierarchy-wont-tell-you-about-motivation

Freud, S. (1895). *Studies in Hysteria.*

Glenn Ballard, I. T. (2016). *Target Value Delivery, Practitioner Guidebook to Implementation.* (K. C. Kristin Hill, Ed.) Arlington, VA: Lean Construction Institute.

Greggory Howell, H. M. (2005). *The Last Planner System: Conversations that Design and Activate The Network of Commitments.* Arlington, VA: Lean Project Consulting.

Harvard Business Review. (2012). *Guide to Project Management.* Boston, MA: Harvard Business Review Press.

Herzberg, F. (1973). *Work and the Nature of Man.* New York: New American Library.

Jeff Sutherland, E. A. (2017). *Manifesto for Agile Software Development.* Retrieved from Agile Manefesto : http://agilemanifesto.org/

Joomis, D. M. (2007). *Building Teachers: A Constructivist Approach to Introducing Education.* Belmont, CA: Wadsworth.

Kristen Hill, C. C. (2016). *Target Value Design - Practitioner Guidebook to Implementation.* Arlington, VA: Lean Construction Institute.

Healthcare Information Technology In Motion

Layton, M. C. (2012). *Agile Project Management - for Dummies*. Hoboken, NJ: John Wiley & Sons, Inc.

LCI. (2017, Sept 13). *Glossary*. Retrieved from Lean Construction Institute: https://www.leanconstruction.org/learning/educatio n/glossary/

LCI. (2017, October). *Lean Construction Institute, Lean Articles*. Retrieved from leanconstruction.org: www.leanconstruction.org/learning/lean-articles/

Lean Enterprise Institute. (2017, August). *Lean Lexicon 5th Edition*. Cambridge, MA: Lean Enterprise Institute. Retrieved from Lean Enterprise Institute: https://www.lean.org/search/?sc=true+north

Liker, J. K. (2012). *The Toyota Way to Lean Leadership*. New York: McGraw-Hill.

McChrystal, S. (2015). *Team of Teams*. New York: Penguin.

NASA. (2017, Sep 13). *The Lean Project Delivery System - An Introduction*. Retrieved from NASA.gov: https://www.nasa.gov/pdf/293166main_56397main _gregory_howell_forum4.pdf

Ohno, T. (1988). *Toyota Production System: Beyond Large-Scale Production*. Portland, Oregon: Productivity Press.

Ortiz, V. (2017, Aug 30). *5 Levels of the Last Planner System*. Retrieved from Lean Construction Blog: leanconstructionblog.com/5-Levels-of-the-Last-Planner-System-Should-Can-Will-Did-and-Learn.html

Portny, S. E. (2013). *Project Management for Dummies - UK Edition.* Hoboken, NJ: John Wiley & Sons.

Pothitos, A. (2016, October 31). *The History of the Smartphone.* Retrieved from Mobile Industry Review: https://www.mobileindustryreview.com/2016/10/the-history-of-the-smartphone.html

Project Management Institute. (2013). *PMBOK Guide - Fifth Edition.* Newton Square, PA: Project Management Institute.

Schwalbe, K. (2011). *Information Technology Project Management - Revised 6e.* Boston, MA: Course Technology.

Seed, E. A. (2017). *Transforming Design and Construction: A Framework For Change.* (W. Seed, Ed.) Arlington, VA: Signature Book Printing (LCI).

Shook, J. (2008). *Managing to Learn.* Cambridge, MA: Lean Enterprise Institute.

Trotter, F. (2013). *Hacking Healthcare.* Sebastopol, CA: O'Reilly Media.

Uhlman, F. T. (2012). *Hacking Healthcare.* Sebastopol, CA: O'Rielly Media.

Walton, M. (1986). *The Deming Management Method.* New York: Perigee.

Webster Dictionary. (2017). *Word of the Day.* Retrieved from Merriam-Webster: www.merriam-webster.com/dictionary/vade%20mecum

Womack, J. (2003). *Lean Thinking.* New York: Free Press.

Healthcare Information Technology In Motion

Healthcare Information Technology
Integrated Project Delivery

Womack, J. P. (2007). *The Machine That Changed The World.* New York, NY: Free Press - Simon and Schuster.

END

Healthcare Information Technology
Integrated Project Delivery

Healthcare Information Technology
Integrated Project Delivery

www.ingramcontent.com/pod-product-compliance
Lightning Source LLC
Chambersburg PA
CBHW022026170526
45157CB00003B/1368